T0313365

The Business Plan Reference Manual for IT Businesses

RIVER PUBLISHERS SERIES IN MULTI BUSINESS MODEL INNOVATION, TECHNOLOGIES AND SUSTAINABLE BUSINESS

Series Editors:

PETER LINDGREN
Aarhus University
Denmark

ANNABETH AAGAARD
Aarhus University
Denmark

Indexing: All books published in this series are submitted to the Web of Science Book Citation Index (BkCI), to CrossRef and to Google Scholar.

The River Publishers Series in Multi Business Model Innovation, Technologies and Sustainable Business includes the theory and use of multi business model innovation, technologies and sustainability involving typologies, ontologies, innovation methods and tools for multi business models, and sustainable business and sustainable innovation. The series cover cross technology business modeling, cross functional business models, network based business modeling, Green Business Models, Social Business Models, Global Business Models, Multi Business Model Innovation, interdisciplinary business model innovation. Strategic Business Model Innovation, Business Model Innovation Leadership and Management, technologies and software for supporting multi business modeling, Multi business modeling and strategic multi business modeling in different physical, digital and virtual worlds and sensing business models. Furthermore the series includes sustainable business models, sustainable & social innovation, CSR & sustainability in businesses and social entrepreneurship.

Key topics of the book series include:

- Multi business models
- Network based business models
- Open and closed business models
- Multi Business Model eco systems
- Global Business Models
- Multi Business model Innovation Leadership and Management
- Multi Business Model innovation models, methods and tools
- Sensing Multi Business Models
- Sustainable business models
- Sustainability & CSR in businesses
- Sustainable & social innovation
- Social entrepreneurship and -intrapreneurship

For a list of other books in this series, visit www.riverpublishers.com

The Business Plan Reference Manual for IT Businesses

Fernando Almeida

University of Porto and INESC TEC

Portugal

José Duarte Santos

ISPGaya and ISCAP

Portugal

LONDON AND NEW YORK

Published 2018 by River Publishers
River Publishers
Alsbjergvej 10, 9260 Gistrup, Denmark
www.riverpublishers.com

Distributed exclusively by Routledge
4 Park Square, Milton Park, Abingdon, Oxon OX14 4RN
605 Third Avenue, New York, NY 10017, USA

The Business Plan Reference Manual for IT Businesses / by Fernando Almeida, Jose Duarte Santos.

Routledge is an imprint of the Taylor & Francis Group, an informa business

ISBN 978-87-7022-039-2 (print)

While every effort is made to provide dependable information, the publisher, authors, and editors cannot be held responsible for any errors or omissions.

Contents

Preface

In order to understand the process of writing a business plan, students must develop multidisciplinary skills that allow the application of knowledge from different fields, such as management, innovation, and technology. In the creation of a new technological company in the IT sector, all these competences are fundamental, together with communication skills to achieve group success. This book presents an innovative template proposal for a business plan and provides a reference manual for undergraduate and graduate students that intend to launch their start-up business in the IT field. It will help them to create and model the business plan of their business. Therefore, this book is mainly aimed at instructors who want to offer a practical view of the modeling process, designing and developing an IT start-up.

The process of writing this book was developed in 2017, from the difficulties experienced by the students in writing a business plan. Over the last 10 years, we have been involved in the development of new business proposals in an academic context for the information technology sector. Throughout these years, we have verified the existence of several books and template proposals for writing a business plan, but which in most situations is not a facilitator of working in multidisciplinary teams made up of students of management and computer science. The vast majority of books available on the market essentially address this issue from the conceptual point of view and suggest a relatively general structure for a business plan that could be adopted across different sectors of activity. However, it has been found that this generic structure does not meet the specificities of several technical areas, in particular the information technology sector.

It is suggested that students may be divided into working groups (e.g., 4–6 students), in which it is advisable to involve students with multidisciplinary competences (e.g., IT and management skills). Each group can use this book as a tool to follow up and answer their doubts. It is also recommended that students may be involved in the creation of the business plan and a prototype. "Chapter 5 – Prototype Description" suggests the use of the UML language in the process of modeling requirements and architecture of the prototype. Additionally, the adoption of prototype software (e.g., Mockplus, Balsamiq, and InVision) as a prototype modeling tool is recommended.

List of Figures

List of Tables

List of Abbreviations

AFR	Annualized Failure Rate
APAC	Asia and Pacific
COD	Cash On Delivery
CORBA	Common Object Request Broker Architecture
CRM	Customer Relationship Management
CSP	Cloud Service Provider
DHCP	Dynamic Host Configuration Protocol
DNS	Domain Name System
EMEA	Europe, Middle East and Africa
EPS	Earnings Per Share
GPL	GNU General Public License
GPM	Gross Profit Margin
ICT	Information Communication Technologies
INPI	National Institute of Intellectual Property
IPR	Intellectual Property Rights
IRR	Internal Rate of Return
ISMS	Information Security Management System
IT	Information Technology
KPI	Key Performance Indicator
LGPL	GNU Lesser General Public License
MTBF	Mean Time Between Failures
MTTR	Mean Time To Recover
NAT	Network Address Translation
NCTAA	National Center of Tourist Animation Agencies
NPV	Net Present Value
OMG	Object Management Group
OS	Organizational Structure
P2P	Peer-To-Peer
PEST	Political, Economic, Socio-Cultural and Technological
PESTEL	Political, Economic, Socio-Cultural, Technological, Environmental and Legal

QoS	Quality of Service
R&D	Research & development
ROE	Return On Equity
ROI	Return On investment
SMB	Small and Medium-Sized Businesses
SOA	Service-Oriented Architectures
TCO	Total Cost of Ownership
TDA	Debt to Total Assets
UCoIP	Unified Communications over IP
UML	Unified Modeling Language
VLAN	Virtual Local Area Network
VM	Virtual Machine
WAB	Web Accessibility Barrier
WAQM	Web Accessibility Quality Metric
WPF	Windows Presentation Foundation
WRL	Workload Rate Limit
WTTC	The World Travel & Tourism Council

Introduction and Contextualization

Objectives

Currently, there are several reference books available on the market that intends to guide an entrepreneur in the process of creating a business plan. However, the vast majority of them essentially addresses this issue from the conceptual point of view and suggests a relatively general structure for a business plan that could be adopted across different sectors of activity. Despite that, it has been found that this generic structure does not meet the specificities of several technical areas, in particular the information technology sector.

In this sense, the purpose of this book is to provide a technical and scientific reference manual for entrepreneurs who intend to launch their business project in the information technology sector. Throughout the book, the concepts about business plan are presented and accomplished in practice through two case studies. This book can also be perfectly adopted in the context of university and professional education levels as a reference guide that could assist the work performed by students in the context of curricular units in the field of entrepreneurship and/or innovation.

What Is a Business Plan?

The business plan is the most relevant document that structures a business project. It allows analyzing the feasibility of a considered business and constitutes the basis of presentation of the project to third parties. Its existence allows reducing the risks and uncertainties for the entrepreneur, company, or investors.

A business plan allows the entrepreneur to organize all important elements of a new business since the moment it is created. In order to be more effective, we must emphasize the information that is more relevant to the business and present it in more detail to external entities, such as risk capitals, banks, suppliers, and potential partners. The business plan should act

1

as a guide in the process of launching a new business, and can be consulted and updated throughout the company's activity. In this sense, the business plan should act as a valuable tool to support management.

A business plan presents itself as an essential tool for several reasons:

- It is a process of validating an idea, through which the entrepreneur obtains elements to decide, from a strategic and financial point of view, whether or not to start a new business;
- By creating a business plan, the entrepreneur studies the viability of his business, creates preventive actions against possible threats and challenges, and analyzes the market and potential clients in depth, avoiding unnecessary efforts and unprofitable investments;
- It is a mechanism that the entrepreneur has to reflect on the potential success of his business;
- It is also a means of internal communication between an entrepreneur and his team;
- It is a dynamic management support tool. Whenever necessary, the document should be adjusted and revised according to the evolution of business requirements.

Business Plan Structure

The basic structure of a business plan is quite flexible. However, some elements are necessarily always important. For example, it is important to know the business model, market share, competitors, amounts of investment, sources of funding, etc.

It is a consensus that a good business plan should be concise and succinct. Therefore, it should not contain more than 50 pages, having in all cases an "Executive Summary" section composed of a unique page. A good business plan should be easy to read and contain tables, graphics, and charts to guide the reader and provide a general idea of the main content.

Our proposal in this book is to adopt a structure composed of six modules:

- Executive summary;
- Chapter 1 – Business concepts;
- Chapter 2 – Marketing plan;
- Chapter 3 – Operational plan;
- Chapter 4 – Financial plan and viability analysis;
- Chapter 5 – Prototype description.

This structure has been defined taking into account different approaches to the construction of the business plan, which basically considers a disaggregation of some of these chapters in others smaller (e.g., marketing plan into products/services and market, financial plan into investment plan and economic-financial projections). We chose to aggregate these dimensions into a single chapter, which in our view facilitates the process of analyzing a business plan. It is also relevant to mention the inclusion of "Chapter 5 – Prototype description", which is innovative and intends to take into account the application of this business plan template to the information technology sector.

Scenarios

In order to facilitate the understanding of the various sections of a business plan and to assist the reader to understand the challenges, decisions made, and difficulties in the practical elaboration of the document, two fictitious scenarios were considered. These two cases correspond to two technology-based companies in the field of information technologies:

- *TourMCard* – technology start-up in the tourism sector;
- *AuditExpert* – technology start-up in the field of auditing in information systems.

Executive Summary

The executive summary is the most important section of a business plan. It is traditionally the first section of the business plan that investors will read and may be the last one if it is not well constructed or enough attractive. An executive summary should briefly describe the company, the product(s), or service(s).

The construction of the executive summary is necessarily difficult given its great importance. It needs to be prepared very well in advance to ensure that all the crucial elements are available there. It is wrong to assume that the omission of any element has little relevance, since the information can be found in any of the remaining sections of the business plan. In fact, it must be assumed that potential investors will only be willing to go through the whole document if they consider that the executive summary is sufficiently appealing.

The following essential points should be presented in the executive summary:

- What is the name of the business and its area of activity?
- What is the mission?
- What is the market need?
- What is the scope of the business and the potential market for your products or services?
- What are the innovative and differentiating aspects of your business?
- What resources (e.g., human, material, and financial) are needed?
- What are the strength and weakness points of the business?
- What are the financial prospects, including the expected deadline for the business to begin presenting profits?

Executive Summary (Scenario I – TourMCard)

Travel & Tourism's impact on the economic and social development of a country can be enormous. It offers great potential for trade and capital investment, creating jobs and entrepreneurialism for the workforce and protecting heritage and cultural values. In Portugal, tourism is one of the main sectors with its weight in the economy growing in recent years.

The intention of *TourMCard* is to improve services in the tourism sector. We have created a tourist card that gives discounts to essential services in holiday destination. We intend to form partnerships with the most relevant players. These partnerships will be developed through direct contact with each partner, including hotels, municipal establishments, travel agencies, and public transport, among others, to spread their services and products through advertising on our platform while participating actively in offering discounts through our pack's promotional.

The tourist can access to *TourMCard* in a physical device that can be purchased on our website or at different sales points, which are located at your destination. There is also the possibility to send the card to the tourists' home address, which allows him/her to plan more conveniently the trip before the tourists arrive to the destination. The *TourMCard* associates a single and unique login for each tourist, where he/she can easily customize his/her profile using a Web browser on his laptop, personal computer, or mobile device. The cloud application is Web responsive. Through this login, tourists will have access to pack's services associated with our network of partners, an interactive map for location and identification of the points of interest near. You still have the possibility of acquiring tourist itineraries suited to their profile and rating and reviewing each partner taking services and their experience into account.

For partners, there is also a unique and personalized dashboard area where they can create, manage, and update your information; interact with the *TourMCard* team in order to make changes to your campaigns; request studies of customers or recommendations; rent advertising space on the page; and respond to criticism and comments from tourists. Furthermore, the interaction with tourists allowed us to conduct studies as our associates in order to know the trends, behaviors, profiles customers, benchmarking, and awareness, among others.

With the focus on value creation, we aim to promote innovation in the sector, ensure tourist satisfaction, and support the growth of members. Financial projections allow us to confirm the profitability of the business. The business presents a low value of payback (1 year and 8 months), and the NPV is higher than 250,000 Euros with an IRR of 123.5%.

Executive Summary (Scenario II – AuditExpert)

Organizations operate 24*365 hours per year and are subject of access and modifications by hundreds or thousands of people each day. In fact, the growing dependence of companies, governments, and institutions from ICT processing systems increases the risk of data security loss. Additionally, business drivers for integration with enterprise management systems have led to IT systems for critical infrastructure becoming interconnecting with corporate networks and directly or indirectly connected to the Internet. This high level of integration can extend to remote access by operational staff, suppliers, and external entities, further increasing the expose of these systems to network vulnerabilities associated with Internet threats. Therefore, it urges the implementation of comprehensive information security systems and policies that could measure and minimize the risk of potential damages associated with this threat.

AuditExpert offers an affordable auditing ICT platform that increases security, assists compliance, and increases operational continuity across the entire IT infrastructure, by looking and analyzing the state of the IT infrastructure of a company. *AuditExpert* uses exclusively open source software and offers a reliable and modular architecture that turns easier to adapt to different companies size and uses an efficient process based on non-intrusive agents that can measure the security state of an IT infrastructure. The process covers the analysis and mitigation of the following core ICT areas:

- Confidentiality – identity management, virus detection, and access controls help ensure that data is adequately secured;
- Integrity – data input, processing, interface and output management, and monitoring controls ensure the accuracy and completeness of records;
- Network resilience, backup/restore arrangements, and environmental data center controls help minimize the probability and impacts of ICT service disruptions such as hardware/software failures or deletion by error or intentionally.

AuditExpert offers also detailed report information as a premium service. In this situation, qualified and credential technician visits each company and examines in detail the ICT infrastructure and organization on the basis of documentation and interviews with key representatives of the organization. The analysis employs the COBIT methodology. The result of the ICT infrastructure audit is a report containing a detailed evaluation of the effectiveness and safety assessment of ICT infrastructure, compliance with ICT best practices, including the practices identified by the standard PN-ISO/IEC 17799:2003, and an assessment of infrastructure management processes. The report can optionally include guidelines for the organizational solutions that enable the customer to obtain in the future ISO 27001 certificate. In addition, the report may contain proposed changes to the ICT infrastructure management model, in accordance with best practices and standards such as ITIL, ISO 20000, and COBIT reference model.

The business presents financial profitability in all considered scenarios. In the standard scenario, we may estimate an NPV of more than 200,000 Euros and an IRR of 58.94%. The investment made in the business is recovered in an estimated period of 3 years and 9 months.

1

Business Concepts

Overview

The purpose of business concepts is to summarize and introduce the context, objectives, and organization of the company. We start to describe the business idea, registration details, and business presentation, which includes the mission, vision, value proposition and organizational values, business premises, organization chart, management and ownership, key personnel, innovation strategy, risk management, and legal considerations. We finalize by presenting these elements applied to the *TourMCard* and *AuditExpert* companies.

1.1 Theoretical Foundations

1.1.1 Business Details/Idea Description

The idea of creating a business can come in many forms, such as discovering a new need, revealing a potential new use for an existing product, new way of distributing a product, or a new form to produce a product.

Regardless of the origin of the idea for the creation of a new business, there is the need to describe it objectively and incisively. One of the first distinctions to make is whether the company will focus on services, products, or both, thus covering two areas of business with different needs in the latter situation.

To describe the business idea, here are some questions to consider:

- What customers' needs intend the idea to satisfy?
- What products and/or services will be marketed?
- What services will be aggregated to the products?
- What are the innovative features of the business?
- What are the competitive advantages compared to existing solutions?
- How will the company be positioned in the distribution circuit?

- What is the expected evolution of the market where the business idea addresses?

In the information technology sector, technological evolution is a constant and one of the main challenges in establishing a new business idea. New business ideas appear constantly and need to be rapidly explored, so it is highly recommended to be aware of fairs and exhibitions where new gadgets and trends are presented. Most promising business ideas in the IT field include:

- Web design;
- Social media consulting;
- Video production and blogging services;
- Data analytical businesses;
- Smart manufacturing and Internet of Things;
- Outsourced call center;
- Internet provider;
- Selling software applications;
- Cloud computing;
- Mobile phones and accessories.

1.1.2 Registration Details

The process of choosing the name of the business should not be considered lightly, as there are a number of limitations, standing out the fact that it has to be a non-existent name in the sector of activity in which it is going to operate. The commercial name should be noted, which is often only composed of one or two words taken from a more complex designation, which facilitates marketing communication.

Ideally, the name is not associated with a product type, because although the customers' needs may be maintained, the way it is satisfied may change. On the other hand, it is also necessary to take into account aspects related to internationalization and to ensure that the name is available in the desired destination countries and that it has no negative connotation. The possibility of registering the Web domain with the chosen name is also a factor to take into account at the time of the decision.

There are several options to find a name:

- Patronymic: name of the founder (e.g., Ford);
- Acronym: results from a company name that has become an acronym (without special sense). For example, IBM (International Business Machine);

- Evocative: the brand name remembers the product category it identifies (e.g., Microsoft – Software for Microcomputers);
- Fantasy mark: may have a previous meaning (e.g., Apple).

The process of registering a business varies from country to country. It is necessary to take into account the requirements, the stages of the process, the time inherent to each phase, the necessary documents, and the value involved, for example, procedural costs. In some countries, such as the United States, it is also important to consider the region in which the registration is carried out, as there are variations depending on the state (including taxes collected), and it may be more favorable to initiate the process in a specific state.

When registering the business, the legal structure has to be indicated, which will support it, considering that there are nuances from country to country. However, we can emphasize the following structures:

- Solo trader: type of business entity which is owned and run by one individual and where there is no legal distinction between the owner and the business;
- Partnership: as two or more persons in business with a unified view to making profits; the number is usually limited to a maximum of 20. Partnership would have the advantage of being able to raise more money because each partner could make a financial contribution. On the other hand, with partners, it is possible have more and diverse competences;
- Company: company's finances are separate from the personal finances of their owners. Having no limit to the number of shareholders it can have, it may only sell its shares privately and it is therefore restricted in the amount of capital it can raise;
- Trust – an entity holding the assets of the business for the benefit of others.

In certain countries, taxes and fees that will affect the business may change according to the legal structure, so it is advised that this aspect is also considered at the moment of decision making by the form of entity.

1.1.3 Business Presentation

Creation of a company presupposes that there is a contextualization of the scope of the business in which it will operate. Considering strategic planning, the first step is based on the definition of vision, which is often thought nebulously in the minds of the founders of the company. Thus, we must first write it, in a future-oriented way, seeking to extol the special contribution that the company can develop for humanity. It translates comprehensively a set of intentions and aspirations for the future, without designating how to achieve it. The idea is that the vision be transcendental to time, a source of inspiration for the collaborators, a reflection of a constant search for a unique event, the search for a dream, the scope of which can be extended to a continuous extension of an eternal perfection. As an example, we can mention the vision of the company Apple, in whose genesis was its contribution to the development of an incredibly fantastic computer.

The next step is to define the mission, which is translated in the so-called mission statement of the company, which should show the reason for existence of the organization, being focused on the market and not on the products. Mission is a broad and enduring statement of purpose, which emphasizes the organization and distinguishes your business from competitors, delimiting your activities in the market you want to occupy. It targets a number of target audiences, with emphasis on customers, business partners, and employees, thus encompassing external and internal perspectives. There is a frequent temptation to enter words such as leader, quality, and customer needs, which should be avoided.

The company to achieve its mission is governed by a set of principles, which are often identified in the mission statement itself. Organizational values are orientations of the external and internal behavior that will be reflected in three organizational levels: strategic, tactical, and operational. They is a contribution to the formation of a corporate identity, which must be defined objectively and in an advisable number between four and six.

The value proposition that the company seeks to transmit to the market must show the reason for the customer to buy from our company and not from the competition. Customers buy from the company, which according to their perception offers them more value. This value should distinguish the company from the competitors and can be constructed by keeping in mind the entire experience of the customer with the product or service, i.e., before, during, and after purchase. Nowadays, everything contributes to the customer experience of the brand, whether through physical contact or online.

1.1.4 Business Premises

The inclusion of the company into a technology park can and should be considered, especially if the company bets strategically on innovation. Therefore, a prior survey of technology parks in the country and in the region sought, if there is already a concrete idea, is important in analyzing what infrastructures, companies, research centers, and existing universities can leverage the business in question.

A technology park, also designed in several countries as science parks, is essential for many IT businesses. Technology parks are close to major universities in the region, as well as other national and municipal public entities. This closeness can produce synergy, enabling the maintenance of services, infrastructure and human resources, the development and transfer of technology, and the creation of new businesses.

Participation in an industrial or commercial cluster should be well thought out, with a view to describe the benefits, drawbacks, and difficulties of operationalizing it.

The characterization of the various facilities should be mentioned in this section. We can differentiate facilities in various types, taking into account specific needs. There are common aspects that also deserve a reflection, such as location, size, acquisition/rental costs, maintenance costs, and possibility of evolution. Some types of facilities include:

1. Manufacturing plant;
2. Warehouse;
3. Administrative facilities;
4. Points of sale.

An analysis should also be made of the possibility that the first three types of installations are not geographically separated, as there are tendentially more disadvantages than advantages. On the other hand, if we have more than one manufacturing unit, there is a need to decide on centralization of administrative services, dispersion, or need for redundancy. This aspect and others have implications in the organizational structure that should be detailed in the following section.

In retail businesses, there are also similar decisions to make, such as whether there are one or more warehouses, whether a warehouse is attached to a point of sale, or an administrative facility, if the administrative premises are geographically close to some point of sale.

The typology of the point(s) of sale should also be well studied, choosing the most appropriate format taking into account various aspects such as the target market and the type of product. All these decisions must be properly framed in the marketing plan.

Also the geographical location of the facilities should be mentioned in this section, emphasizing the criteria that supported the choice. In case the business is associated with a brand that already has points of sale located in the same geographic region, it is necessary to take into consideration and try to define the relationship.

Ideally, a study should be presented that identifies which areas will be required to exist, such as warehouse technical support and their dimensions.

1.1.5 Organization Chart

The organizational structure (OS) constitutes as a way to operationalizing the idealized strategy. There are different models and all of them can be chosen by the organization, taking into consideration that the organizational structure adopted is the one that is deemed the most appropriate to achieve the proposed objectives, that is, the structure that contributes most to the effectiveness of the organization. However, we must also keep in mind the level of efficiency that will be achieved by the structure chosen. There is no finished, or perfect, OS, but we should be able to adapt to the necessary changes.

Other aspects that condition the design of the OS are:

- Environment: growth capacity, volatility, and complexity;
- Type and level of technology incorporated into the organization;
- Task environment, whose complexity may motivate a division of labor;
- Age and size of the organization;
- Entrepreneurial leadership style, power, and control desired.

It is necessary to make decisions about the size of the organization, the details of departmentalization, and the level of horizontal and vertical differentiation, and this last decision influences the flattening of the organization. It should also be noted that the structure adopted tends to influence the behavior and attitudes of workers.

The structure of the organization should be graphically represented through an organizational chart that portrays departmentalization, existing functional areas, and relationships between them. The structure of the business can be drawn more generally, designating the departments, or in detail, it being usual to refer in this situation what the functions are and also the names of the people who occupy them.

The simplest organizational structure (see Figure 1.1) tends to be common in small family businesses. It usually consists of only two levels: the entrepreneur and the collaborators.

An organizational structure based on the principle of departmentalization tends to use one of the following types:

- By functions;
- By product;

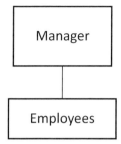

Figure 1.1 Simple organizational structure for small family businesses.

- Per customer;
- By geographical area;
- By division;
- By project;
- Matrix;
- Combining forms;
- Multifunctional teams.

The organizational structure by functions (see Figure 1.2) is the most classical and brings the grouping of people into specialized activities, which carry out activities within the same technical or knowledge area. It is recommended that people in different departments remember to have a global view of the organization, avoiding excessive concentration in their specific area.

In the organizational structure by product (see Figure 1.3), all the people who deal with the product in question are grouped in the same unit. It is recommended in companies that have extensive product ranges or that their specificity is of a complexity that requires specialized knowledge. The failure of a product may call into question the unit to which it is assigned. It is not a very suitable model for sectors of activity where competition is minimal or not very active.

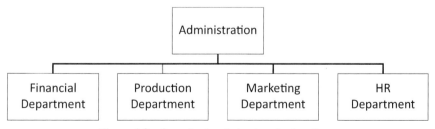

Figure 1.2 Organizational structure by functions.

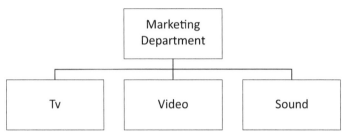

Figure 1.3   Organizational structure by product.

The structure organized by clients (see Figure 1.4) tends to be used by companies in which customer relationships imply very different concerns.

In companies that use a structure organized by geographical area (see Figure 1.5), the activities are organized according to the geographic location where the business activities take place. This model presents a decentralization of the decision-making process.

The division structure (see Figure 1.6) is based on the separation of tasks based on the diversity of the company's products, services, markets, or processes. Each division has its functional structure and may be advisable in companies with a diversification strategy.

In companies that develop large projects or endeavors, the organizational structure by project presents (see Figure 1.7) itself as a solution to be considered. People from different departments with different experiences and competencies are associated for a given period of time on a project.

When the activities of the company no longer have a temporal framework, as in the previous case, the option may be the matrix structure (Figure 1.8).

The organizational structure necessary for the company to be able to operate in the market may have to be more complex and involve the combination of several forms, such as the model presented below (see Figure 1.9), where the structure is present by function, by product, and also by geographical area.

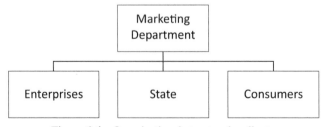

Figure 1.4   Organizational structure by clients.

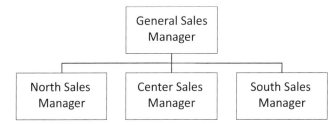

Figure 1.5 Organizational structure by geographical area.

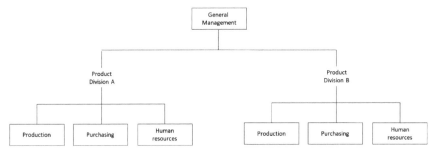

Figure 1.6 Organizational structure by division.

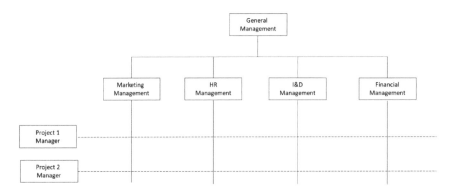

Figure 1.7 Organizational structure by project.

There are also other alternatives (see Figure 1.10) to the traditional departmental organizational structure, standing out the structure organized by teams, multifunctional, autonomously, and tendentionally with more flexibility compared to the previous structures presented.

In start-ups, a business organizational method widely used is the adoption of flat organizations, in which hierarchy is ignored. In this sense, decisions are made collectively, by consensus. The goal of a flat organization is to

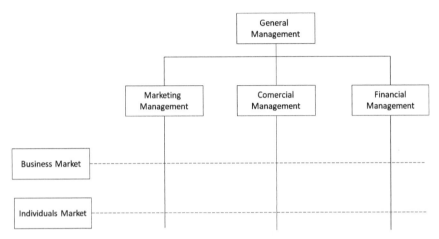

Figure 1.8   Organizational structure by matrix.

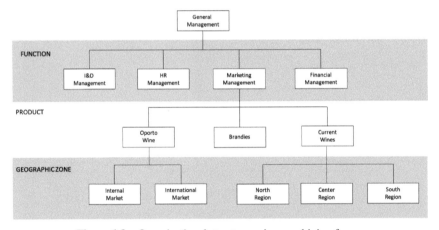

Figure 1.9   Organizational structure using combining forms.

involve all people in the decision-making and execution process. Another advantage of this model is that employees loose less energy with internal disputes. People's energy focuses on contributing to the strengthening and growth of the company and not to hierarchical growth.

1.1.6 Management & Ownership

In this section, we intend to make a brief presentation of the entrepreneurs who will be in the genesis of the business. This presentation should include personal data, such as the name, and also aspects related to academic and

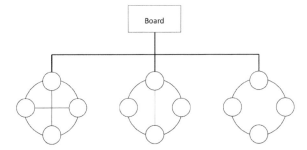

Figure 1.10 Organizational structure using multifunctional teams.

technical training. The personal, social, and professional competences must be well evidenced, as well as the professional background.

It should be kept in mind that the business plan is an internal document, but that in certain circumstances it may be disclosed, for example, in a limited way to obtain funding. In this case, there is a need to convey to prospective investors the profile of the promoters and their degree of commitment to the project.

On the other hand, there is also a need to differentiate the owners who work in the company from situations where owners only play the role of an investor. In this last situation, there is a need to define who will be ahead of business, what are the criteria that will be present in the choice.

When there are investors who are companies, there is a need to keep in mind the added value they can provide and the role they play in this project.

In summary, it is intended to be mentioned the contribution that each of the partners proposes to provide for the success of the business.

The distribution of the company's capital by each of the investors should be mentioned bearing in mind that there may be minimum values or quotas according to the legal structure and/or the country where the company's head office is located.

1.1.7 Key Personnel

In this section, we want to highlight the key people of the organization, for example, those in charge of the company departments. To facilitate this, we can use the organizational chart presented above.

In the case of an existing company, already having staff, a list must be created composed of the job title, the name of the person occupying this function, the expected staff turnover, the skills, and the knowledge that

Table 1.1   Current staff

Job Title	Name	Expected Staff Turnover	Skills and Strengths

the person occupying that position must have. Thus, the Table 1.1 must be completed.

In the IT field, we typically have very low expected turnover due to the shortage of qualified professionals in this area. Hence, due to the high internal and external competition between companies, it is difficult to retain employees for a long time. Moreover, since the skills of the employees are very specialized, it becomes difficult to go to the market and make a direct hiring without considering the need of time for the training of this new collaborator.

Also a table should be included, where the necessary staff forecasts are listed, according to the following example (Table 1.2).

Recruitment Options

When the company is already operating, there are two major recruitment options: internal and external.

Internal recruitment has lower costs and it is a motivating factor and commitment of employees. When used in excess, it may not allow personnel to be renewed. On the other hand, external recruitment presents as advantages the renewal of competences and the introduction of new ideas. But it is more expensive and there is a possibility that people recruited will not adapt to the company's culture.

When we establish a new company in the market, external recruitment is the only option. In strategic terms, the company must decide if it will privilege a recruitment process within the sector of activity where it operates, if it chooses to only recruit in other sectors, if it is indifferent, or if it depends on the concrete situation. This decision has consequences on the definition of the requirements, or may eventually be postponed to the selection phase. It is also necessary to decide if the disclosure of the recruitment is done anonymously,

Table 1.2   Required staff

Job Title	Quantity	Expected Staff Turnover	Skills and Know-how Necessary	Date Required

which limits the way the recruitment is published. For example, the company's website can no longer be used. Another strategic decision is whether outsourcing or not the recruitment and selection process.

In the IT field, it is common that a company adopts simultaneously several recruitment options. The use of social networks is also a widely adopted practice, and it is equally important to publicize professional offers among recent graduates. It is also common to offer part-time positions for students attending undergraduate and master's degrees.

Training Programs

Regardless of the source of recruitment, a strategic training plan for employees should be defined, identifying priority training areas and defining how to ensure such training (internal, external, hybrid).

We can consider three major types of training needs:

i. Behavioral training, which tends to be transversal to the whole organization, although there may be training programs that may be more oriented to the role that the person plays. Examples: interpersonal relationship, team management, and time management;
ii. General technical training, which, despite being no longer behavioral, may also be important for employees to have access regardless of their role. Examples: computer science, foreign language, and first aid;
iii. Training of a technical and specific nature, oriented toward a function or group of similar functions. Examples: sales techniques, social network management, and labor law.

At the operational level, in detail, a training plan can be presented which includes the identification of the employees, the training programs they are going to attend, the internal persons or the external entity that provides the respective training, place of training, and schedule.

In the IT field, the three major types of training are important and can be used as a retention strategy. It is a relevant complement to the salary of the employee, since the IT professionals consider that have new knowledge and competencies in emergent technologies are very important for them.

Skill Retention Strategies

Retaining talent means working on two dimensions: the person and the organization, that is, acting directly in each one of the employees, taking into account the specifics of each one, simultaneously acting in the whole of the company, dynamizing an organizational culture that goes towards to provide an environment that integrates employees and encourages them to evolve.

In business plan, we should try to answer the following questions:

- Which employee evaluation system does the company intend to own/possess?
- What evaluation system for internal training needs does the company have?
- How is the career plan defined for each employee?
- What are the incentives for professional development?
- What mechanisms does the company have to streamline knowledge management?

1.1.8 Innovation

Research & Development (R&D)/Innovation Activities

Innovation can be a catalyst for the growth and success of your business, and help the enterprise to adapt and grow in the marketplace. Thus, a company must identify sources of information that contributes to the internal development of a culture of innovation, such as:

- Internal sources for the company;
- Other companies belonging to the same group;
- Competitors;
- Customers;
- Consulting companies;
- Suppliers of equipment, materials, components, or software;
- Universities or other institutions of higher education;
- Government research institutes or private non-profit institutions;
- Patents;
- Scientific or professional conferences, meetings, and publications;
- Computational information networks;
- Fairs and products demonstrations.

These sources of information will facilitate the development of innovation activities that are all steps necessary for the development and introduction of technologically new or improved products or processes. In the business plan, we must identify which innovation activities the company will privilege, such as:

- Research and experimental development carried out in the company (internal R&D);
- External realization of R&D activities (external R&D), resulting, for example, from cooperation with other companies;

- Acquisition of R&D services (external R&D);
- Acquisition of equipment and machinery linked to product and process innovation;
- Acquisition of other external technologies linked to product and process innovation;
- Industrial design and other activities prior to the production of technologically new or improved products;
- Acquisition of other external knowledge;
- Direct training linked to technological innovation;
- Marketing activities, such as market introduction of technological innovations.

Intellectual Property Strategy

When starting a business, it is crucial to protect its assets. Intellectual property is one of the main assets that business owners fail to protect and can include:

- Trademark;
- Copyright;
- Brand;
- Domain name;
- Recipes or products created.

Thus, an inventory of the items considered important from the perspective of the intellectual property strategy should be made and, at the same time, define how to protect them, identifying the process and external organisms that should be taken into account.

It is also important to define mechanisms to protect confidential or sensitive information, bearing in mind all the facilities that exist today in reproducing, photographing, or filming.

In the case of existence partners, suppliers, or other entities that have access to information critical to the company's strategy, protection should be provided for undue disclosure.

Software protection can be carried out either by patents or by copyright protection. However, these two possibilities are not always present in all markets, and the use of patents is more common in the United States and copyright protection in Europe. In Europe, there are difficulties in the use of patents since it becomes difficult to comply with the requirement of industriability, that is, the program is not the own process that is produced by a machine in which the software runs. Hence, computer programs are considered as copyright in many countries. For this, software is recognized as an intellectual work of linguistic expression, and it must manifest a certain level of creativity. In this sense, when the creativity is minimal or the expression used is the only one possible for the manifestation of the idea, copyright protection is not allowed.

1.1.9 Risk Management

In this section, we intend to perform a survey to identify possible risks, organized by the likelihood, which could impact the business.

The risk analysis should be carried out based on introspection, and also focusing specifically on the sector of activity where the company operates and also on the generic environment.

In the field of risk, we must include a description of the risk and the potential impact on your business. In the likelihood field, it is categorized as *Highly Unlikely*, *Unlikely*, *Likely,* or *Highly Likely*. On the other hand, in the impact field, the level of the effect is identified as *High*, *Medium,* or *Low*. In the controls column, you define how you can verify the risk materialization. In the next field, a strategy is established to deal with the realization of the risk, that is, what actions will you take to minimize/mitigate the potential risk to the business. Finally, we must specify the person or the department responsible for monitoring the risk (Table 1.3).

1.1.10 Legal Considerations

The legislation differs from country to country, and within the same country there may also be differential situations between states. Legislation can be classified into two groups:

- Generic legislation: transversal to all sectors of activity (e.g., health and safety; packaging and labeling laws; traineeships and funding; and environmental laws);
- Specific legislation: specific to one sector of activity (e.g., health and safety; packaging and labeling laws; traineeships and funding; and environmental laws).

Thus, it is necessary to make a survey of all the legislation that can condition the operation of the organization, but also that it limits the great strategic options.

Table 1.3   Identified risks

Risk	Likelihood	Impact	Controls	Strategy	Responsible

The company must also define how the legislation will be fulfilled by the company. For example, will compliance with health and safety legislation be evaluated by a specialist outsourcing, or using only internal resources?

1.2 Scenario I – TourMCard

1.2.1 Business Details/Idea Description

Tourism is one of the main sectors of the Portuguese economy. Its weight in the economy has grown significantly in recent years. Thus, with the intention of improving the services provided in the tourism sector, we created the *TourMcard* website, which, through the partnerships established with entities related to the sector, offers a wide range of discounts and services. These partnerships will be developed through direct contact with each partner to be understood, housing units, hotel units, municipal establishments, travel agencies, and public transport, among others, to disseminate their services and products through advertising on our platform and at the same time to actively participate in the offer of discounts through our promotional packs.

The interaction of the user with *TourMCard* will allow carrying out studies as our associates in order to know trends, behaviors, clients, and benchmarking, among others.

Through its website, *TourMCard* will allow the creation of a unique login for each tourist where, at an early stage, a small interaction with the page can be customized to your profile. Through this login, the tourist will have access to pack's of services associated to our network of partners, an interactive map for location and identification of points of interest. You will also have the possibility to purchase tourist routes appropriate to your profile and to evaluate and comment on the services of each partner experience.

As for our partners, they will also have a unique login where they can create, manage, and update their information; interact with the team *TourMCard* in order to make changes to your campaigns; request client studies or recommendations; rent space for advertising on the page; and respond to reviews and comments from tourists.

With the commitment to create value, we intend to promote innovation in the sector, ensure tourist satisfaction, and support the growth of members.

1.2.2 Registration Details

In choosing the *TourMCard* brand, it was intended to be as global as possible and to diminish the possibility of having some negative connotation in some eventual country. It was also sought to find an evocative designation, thus expressing the scope of the business.

In countries where *TourMCard* is available, there will be a concern to register the card designation, as well as all packet names created, if circumstances dictate.

In the country of origin, Portugal, the company will be registered as sole trader and will have a registered capital of 500 €.

1.2.3 Business Presentation

Mission
TourMCard facilitates the acquisition and allows access to unique tourism products.

Vision
We believe that tourism products will all be cataloged and can be purchased, customized, and aggregated into a user account.

Value Proposition
We offer tourists in our cities an interactive trip, with access to discounts, where you can easily identify the touristic sites by excellence.

Organizational Values
The organization seeks to reflect in its operations its organizational values:

- Simplicity in access to services;
- Transparency in the products made available;
- Continuous search for the best business partners.

1.2.4 Business Premises

The company's operations will be supported with an office in Portugal based in the Greater Porto area. In order to ensure the growth of the company and taking into account that all employees are expected to be in the same geographic location, the choice is at UPTEC – Science and Technology Park at the University of Porto. It is also expected to benefit from a set of infrastructures that leverage the company's capacity for innovation.

1.2.5 Organization Chart

The graph below shows how you want to structure the company. The functional areas shown in gray do not correspond to the start-up phase, but to the

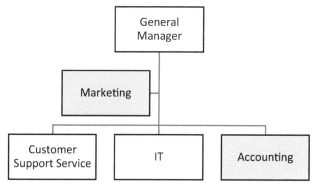

Figure 1.11 Organizational chart of *TourMCard*.

desired future situation. The organization chart illustrates that the company will be organized by functions (Figure 1.11).

1.2.6 Management & Ownership

Management will be carried out by the sole partner and owner of the company at an early stage. However, it should not be ruled out that after the first 5 years of the business, it will be possible to extend ownership to other people who might be considered strategic, such as business partners.

This opening of capital to other investors may also be a possibility, in the event of an internationalization involving direct investments in the countries of destination.

1.2.7 Key Personnel

Table 1.4 presents the employees who will be present at the beginning of the company. The number presented is considered as the ideal value to launch the business, covering the areas crucial for the business. On the other hand, the project is based on being launched with the lowest possible number of employees, to avoid an initial financial effort that cannot be supported with income or funding.

Table 1.5 shows the forecast requirements taking into account the evolution of the company. Thus, an increase of 2 people in 2018, 3 people in 2019, 2 in 2020 and 2 in 2021, reaching a total of 12, is considered to be feasible.

Recruitment Options

Considering that the company is a start-up with few human resources, the recruitment will have to be carried out externally. However, it cannot be ruled

Table 1.4   Initial staff of *TourMCard*

Job Title	Name	Expected Staff Turnover	Skills and Strengths
General manager	Joseph Sotnas	Never	Relevant qualifications in management; spirit of entrepreneur.
Web developer	Andy Russel	4–5 years	Computer engineering; programming domain; creativity; guidance for business.
Customer support service	Joan White	3–4 years	Sales and customer orientation; knowledge of languages; proactive and assertive.

Table 1.5   Required staff of *TourMCard*

Job Title	Quantity	Expected Staff Turnover	Skills and Know-how Necessary	Date Required
Marketing manager	1	3–5 years	Qualifications in marketing; Internet and strategic marketing specialist; 4 years experience in the function	2018
Web developer	1	4–5 years	Computer engineering; programming domain; creativity; guidance for business	2018
Customer support service	2	3–4 years	Sales and customer orientation; knowledge of languages; proactive and assertive	2019
Web developer	1	4–5 years	Computer engineering; programming domain; creativity; guidance for business	2019
Customer support service	2	3–4 years	Sales and customer orientation; knowledge of languages; proactive and assertive	2020
Accounting	1	6–8 years	Knowledge of accounting and taxation	2021
Marketing assistant	1	3–5 years	Two years experience in services marketing; deep knowledge of Internet marketing.	2021

out that in the first stage, the provision of vocational traineeships can be made and then the trainees can be admitted to the company's staff.

Therefore, the company reaches a dimension that justifies being in charge of the IT department, and customer support service will be privileged internal recruitment.

Training Programs

Training will be a constant concern of the company. In the technological component, we must follow the appearance of new technologies and computer tools. Therefore, every year, the elements that integrate the IT department must improve and develop new knowledge and skills in the area of development, integration, and maintenance of systems that support the activity of the organization.

In the customer support service department, there will be continuous training in customer relationship management, with special focus on platforms that provide interaction, taking into account aspects related to good practices that contribute to the increase of brand awareness.

Crosswise to the company, we will also give special attention to training in interpersonal relationships, seeking to continuously strengthen the team spirit.

Skill Retention Strategies

The company for each of the collaborators will develop a career plan that will contemplate whenever possible ascension to position of superior level to the one performed.

There will be an evaluation system that will be presented to employees every year. This system assumes that the employee makes his own assessment based on the previously known criteria and, after that, presents and discusses his analysis with his/her hierarchical superior.

1.2.8 Innovation

Research & Development (R&D)/Innovation Activities

It is intended that the company be involved in a culture of innovation, from the perspective of not only developing business processes that dynamize automation, but also adding value to the customer, improving the relationship with the customer, and providing an unforgettable customer journey.

Thus, there will be periodic meetings whose main objective is to create a continuous internal dynamic in the search for new forms of interaction with the client and creation of new services.

Intellectual Property Strategy

The company whenever it develops something that is innovative should seek to register using the INPI – National Institute of Intellectual Property, which is the entity that protects and promotes industrial property.

In order to protect internally developed innovation and ensure standard-ization, information will be classified into:

- Confidential (highest level of confidentiality);
- Restricted (medium level of confidentiality);
- Internal use (lowest level of confidentiality);
- Public (everyone can see the information).

1.2.9 Risk Management

Table 1.6 seeks to highlight potential risks inherent in the business.

Table 1.6   Identified risks of *TourMCard*

Risk	Likelihood	Impact	Controls	Strategy	Responsible
Break in tourism demand	Likely	High	Analysis of tourist trends	Development of new products and markets	Management
Aggressive sales tactics by competitors	Likely	Medium	Continuous monitoring of competition price and promotion campaigns	Promotion sales	Marketing department
Card innovation by competition	Unlikely	Medium	Continuous monitoring of competition products	Innovation in the product	Marketing department
Breach of partnerships	Unlikely	High	Analysis of the satisfaction of partners	Continuous trading	Management
Escape from human resources	Unlikely	Medium	RH turnover	Retention programs	Management

1.2.10 Legal Consideration

The company also complies with all legal requirements taking into account the licensing of companies that organize, offer for sale or sell tourist trips, or broker the sale and booking of other tourist services.

It should also be included in the National Center of Tourist Animation Agencies (NCTAA), integrated in the National Tourism Registry, which is an electronic platform that gathers and makes available information on tourist tourism companies and maritime tourism operators operating in Portugal.

Regarding to legislation, it is necessary to consider:

- Decree-Law No. 209/97 of 13 August;
- Directive 90/314/EEC of 13 June 1990, Ordinance No. 784/93 of 6 September;
- Decree-Law No. 12/99 of 11 January, amending Decree-Law No. 209/97 of August;
- Decree-Law No. 263/2007 of 20 July, which proceeds to the third amendment to Decree-Law No. 209/97 of August.

1.3 Scenario II – AuditExpert

1.3.1 Business Details/Idea Description

Given the increasing value of information, based on the construction of information technology (IT) systems and information communication technologies (ICT), it is essential that appropriate protections are established that will be evaluated or recommended by the security audit.

The performance of the audit of information systems has to follow the different lines of evolution of the various technologies that are included in the field of computer operations. It is necessary to assess whether the security models agree and are in line with the new architectures, the different platforms and the forms of communication, since it cannot be audited with concepts, techniques, or recommendations that have become obsolete.

In this sense, the scope of *AuditExpert* is to provide highly specialized IT consulting services, based on two main pillars: the domain of existing and emerging technologies and the knowledge of the client's business.

The main customers will be medium-sized companies, regardless of the sector of activity. We consider potential customers those that need to monetize or improve their information systems. In the first phase, the business will focus only on the country of origin.

1.3.2 Registration Details

Although in a first phase the company intends to operate only in the Portuguese market, it is intended that the company's name be a factor that drives the internationalization, and therefore the expression chosen has a meaning in the English language. Thus, in a first phase, the registration will be carried out in the country of origin and the establishment of the company as a sole trader and with a share capital of 2000 €.

As internationalization arises, there will be a concern to register the *AuditExpert* brand whenever possible in the concerned country.

1.3.3 Business Presentation

Mission
AuditExpert's mission is to audit information systems, thus contributing to the improvement of efficiency and efficiency of processes and optimization of the business of its clients.

Vision
Be a leader in the provision of technology audit services.

Value Proposition
AuditExpert provides computer audit services that allow your company to increase the security of information and the reliability of your data, Web applications, and mobile platforms.

Organizational Values
AuditExpress's performance reflects its organizational values:

- Confidentiality;
- Proactive attitude;
- Constant evolution;
- Valuing tacit and explicit knowledge.

1.3.4 Business Premises

The company will have its headquarters in Lisbon, in a science park, which is expected to be inspiring for the dynamics of the company that it intends to instill as well as for innovation capacity. The company's operations took place throughout the country at the customers' premises, whenever necessary.

The facilities do not require any special requirements. However, it is crucial that there is access to state-of-the-art technology infrastructures, including access to 5G services as they become available.

1.3.5 Organization Chart

The company will present the simplest possible structure (see Figure 1.12), made up of the management and a department that will integrate the elements

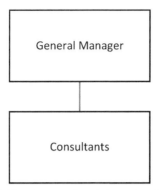

Figure 1.12   Organizational chart of *AuditExpert*.

that will provide the consultancy, but also has commercial responsibility. The marketing will be assured by the management of the company. There will be a concern in subcontracting all services not considered strategic, such as accounting.

1.3.6 Management & Ownership

The company has three partners, one with a 50% share and the other two with a 25% share each. The partner with the largest share assumes global management. The other two partners perform functions in the company as consultants.

The planned expansion into international markets may entail the opening of the capital to other partners as a counterpart of access to these markets.

1.3.7 Key Personnel

The company starts with three employees who are also owners (Table 1.7). The person who occupies the role of General Manager will also perform, when necessary, IT consultant functions.

With the planned expansion for international markets, it is estimated that in each of these markets, there will be a need to exist at least one person who develops the entire business from a commercial point of view but who also has the skills to provide the consulting service sought by the client. Thus, Table 1.8 incorporates this desired future reality.

Recruitment Options

Recruitment in the national market will privilege internships supported in protocols carried out with universities with IT courses. For the company to

Table 1.7   Key personnel of *AuditExpert*

Job Title	Name	Expected Staff Turnover	Skills and Strengths
General Manager	Lydia Davis	Never	Relevant qualifications in management; spirit of entrepreneur; IT consultant senior experience
IT consultant	Simon Walsh	Never	IT consultant senior experience; dynamic and proactive
IT consultant	Julius Rock	Never	IT consultant senior experience; dynamic and proactive

Table 1.8   Required staff of *AuditExpert*

Job Title	Quantity	Expected Staff Turnover	Skills and Know-how Necessary	Date Required
Country partner	1 per country	10 years	In-depth knowledge of the destination country's market; management skills IT consulting experience	2019

operate in the international market and find collaborators with the necessary specificities, there will be recourse to the use of head hunters who will look for candidates who are prospective as future investors.

Training Programs

The IT field has been constantly evolving and accelerating. Therefore, for the company to be competitive, there is a need to ensure internal technical training in the areas of service delivery, as well as in other potential areas to expand the service. The training programs will be defined keeping in mind the process of performance evaluation.

It is important to use software and hardware suppliers to obtain information and training, as well as attend national and international seminars that must be identified each year and be integrated into the annual training plan.

Skill Retention Strategies

The training programs will contribute to retain employees, but there will also be a special concern for the work environment, which is desired family and informal, participatory management, evolution of remuneration, and a constant concern with the principle of equity.

1.3.8 Innovation

Research & Development (R&D)/Innovation Activities

The company will seek to develop partnerships with other entities in the area of hardware and software, in order to participate in innovation projects that allow them to be at the forefront of new technologies, new solutions, and new working methodologies. It will also be important to take advantage of any funded programs, which will overcome possible financial constraints.

Intellectual Property Strategy

With the innovative methodologies or processes that have been developed within the scope of *AuditExpert* activity, there will be a concern to publish articles in specialized journals. It seeks to obtain recognition by the community and simultaneously claims the authorship of the elements in question.

1.3.9 Risk Management

Table 1.9 seeks to highlight potential risks inherent in the business.

Table 1.9 Identified risks of *AuditExpert*

Risk	Likelihood	Impact	Controls	Strategy	Responsible
Cloud expense services	Likely	Medium	Continuous monitoring of services offers	Develop new services	Management
New services of competition	Likely	Low	Competition analysis	Differentiation	Management
Entry of international competitors	Unlikely	Medium	Continuous monitoring of competition	Deepen the knowledge of the domestic market	Management
Merger of competitors	Unlikely	Medium	Merger of competitors	Monetize the segments served by competition	Management
Recession	Unlikely	Medium	Economic environment	Costs optimization	Management

1.3.10 Legal Consideration

The company in its activity will take into account all the legislation, or standards that exist and are related to its existence. In addition, they are interconnected to the products marketed, or services made available, namely Administrative Rule no. 302/2016, of 2 December that amended Administrative Rule 321-A/2007, of 26 March, regarding the data structure of the SAF-T (PT) file; standards ISO/IEC 15504, ISO/IEC 14764, ISO/IEC 14598.

2

Marketing Plan

Overview

The purpose of a marketing plan is to clearly identify and show what steps or actions will be taken to achieve the plan goals of a business. In fact, an effective marketing plan must set clear objectives that will help the entrepreneurs toward the established long-term strategic goals. We start to present the theoretical foundations that guide our readers toward the several components of a marketing plan, such as market research, environment analysis, analysis of customers, partners and competitors, strategic positioning, advertising & sales, marketing mix, and business model canvas. After that, we use two scenarios (*TourMCard* and *AuditExpert*) to demonstrate the application of the marketing plan.

2.1 Theoretical Foundations

2.1.1 Market Research

Marketing research is an elementary section of a business plan. It is used to help in improving management decision-making by providing relevant, accurate, and timely information. In fact, understanding customers and identifying who they are, what they want in terms of products or services, how and where they want it to be available and delivered, and at what price they are available to pay are some of the most important decision criteria a manager must be aware of.

Marketing research can help organizations in various decision-making processes, which can be divided into two groups: (a) problem identification research and (b) problem-solving research. The problem identification research is used to understand and identify problems that are not necessarily visible at a first glance, and problem-solving research is used to help

35

solve specific research problems. Major techniques associated with problem identification research are:

- Market share research;
- Market potential research;
- Sales analysis research;
- Forecasting and trends research;
- Branding and image research.

On the other side, associated with problem solving research, we have the following techniques:

- Market segmentation research;
- Product research;
- Pricing research;
- Promotion research;
- Distribution and logistics research.

One of the most accessible and basic techniques to perform a marketing research is conducting an exploratory research. This technique is adopted to get an initial clarification and definition of the nature of a problem. It is used to diagnose a situation, screen alternatives, and discover new ideas. Typically, it does not provide a full conclusive evidence and subsequent or complementary research is needed. The exploratory research can be based on both qualitative and/or quantitative data.

Another technique that can be used is called descriptive research, which is usually characterized as being concerned with measuring or estimating the sizes, quantities, or frequencies of things. It can be used to measure market size, market structure, and the behavior and attitudes of consumers in the marketplace. Like exploratory research technique, it can adopt different qualitative and quantitative approaches. The most common approaches applied to marketing research include surveys, case studies, field experiments, simulation, and in-depth interviews. In the context of this book, we will look in detail into the survey approach that is adopted in many marketing research studies.

A survey has the potential, when properly conducted, to allow the generalization of beliefs and opinions of a given population by studying a subset of them. It is based on statistics field, particularly in the subset of statistical inference. A survey needs to follow strict procedures in order to let the generalization of data. Several authors define their own stage process, which basically can be grouped into the following five steps:

1. Identify research objectives – present the objectives of the survey, the problem to be addressed and how the study will answer questions about the problem;
2. Identify and characterize target audience – analyze the dimension of the target audience and ensure that those who respond to the survey are representative of the target audience. Guarantee that the target audience is willing to participate in the survey and they understand the used terminology;
3. Design and write questionnaire – define the list of questions and its structure. Choose the format of the questionnaire, for instance, using online surveys. Before distributing the questionnaire guarantee that it was previously tested by some members of the target audience in order to detect some issues and improve it;
4. Distribute the questionnaire – define how the questionnaire will be distributed to the target audience. Identify some key persons that can help in this process;
5. Analyze results and write conclusions – use statistical problems to organize collected data, identify potential correlated variables, and plot graphical representations of the data. Finally, include the relevant information in the business plan and attach a copy of the used questionnaire to the back of the business plan.

Finally, it is important to consider that the Internet is a great tool for marketing research, particularly for IT businesses. It enables primary market research to become much less expensive to conduct than by using traditional media. The Internet can be used to perform exploratory research by reading reports from consultancy firms, thesis, marketing analysis papers, etc., and it can also be used to in the context of descriptive research, for instance, using online questionnaires. However, the Internet does not replace a greater contact of entrepreneurs with their target when it becomes possible, namely by conducting interviews with potential customers, partners, and suppliers, and also with national government and regional and local agencies in order to assess the potential, limitations, and barriers of the business idea.

2.1.2 Environmental/Industry Analysis

Environmental analysis is a strategic tool used to identify all the external elements that may affect the organization. This analysis helps align organization's strategy with the firm's environment.

The environmental analysis entails four steps as described below:

- Scanning – identifying early signals of environmental changes and trends;
- Monitoring – detecting meaning through observations of changes and trends;
- Forecasting – developing projections of outcomes based on changes and trends;
- Assessing – determining timing and importance of changes and trends for firm's strategies and management.

There are many strategic analysis tools that a firm can use. The most common tools include the PEST analysis and Porter's Five Forces.

The PEST analysis is the most common approach for considering the external business environment. PEST analysis stands for political, economic, social, and technological analysis and describes a framework of macro-environmental factors used in the environmental scanning component of strategic management, but the word PEST is no more than a convenient mnemonic. The underlying thinking of the PEST analysis is that the enterprise has to react to changes in its external environment. This reflects the idea that strategy requires a fit between capabilities and the external environment and so it is necessary for an enterprise to react to changes.

The **political factors** look to the country's current political situation. It also includes the global political condition's effect on the country and business. Some common political factors may be the political stability, legal framework, intellectual property rights (IPR) protection, tax policy, and market regulation.

The **economic factors** involve all the determinants of the economy and its state. These are factors that can conclude the direction in which the economy might move and, therefore, it helps the company to set up strategies in line with the changes. Typical economic factors include economic growth, interest rates, inflation rate, exchange rate and currency stability, unemployment rate, credit accessibility, and investment opportunity.

The **social factors** reflect the distinctive mindset of every country. These social factors may have an impact on the businesses by affecting the sales of products and services. Typical social factors include literacy of the place, lifestyle and social trends, demographic characteristics (growth ration, sex ratio, age distribution, population density, etc.), human development index, and social safety and benefits.

The **technological factors** cover the effects of technological change on products, processes, and distribution channels. In fact, technology is changing every day and, therefore, the company should stay up to date with these changes. Some technological factors may be the R&D activity, innovation, licensing, patent acquiring, maturity of technology, skilled resources, easier acceptance of new technologies, and information and communication technology (ICT) technologies.

There are some variations of the PEST analysis that typically extends the level of profundity and organization of the traditional paradigm of PEST analysis. One of the most popular extensions is the PESTLE or PESTEL analysis that adds two other factors (environmental and legal). Others uncommon extensions are:

- PESTELI = PESTEL + industry analysis;
- STEEP = PEST + Ethical;
- STEEPLED = PEST + Environment + Legal + Ethical + Demographic;
- LONGPEST = Local + National + Global factors + PEST.

One of the most important tools to assess the industry attractiveness is known as "Porter's Five Forces". It was proposed by Porter (1998) and aims to make clear how each stakeholder influences, positively or negatively, the business. The five forces are: rivalry among competitors, bargaining power of suppliers, customers' bargaining power, threat of new entrants, and the threat of substitute products.

The **threat of new entrants dimension** determines the difficulty and opportunity for companies that intend to enter the sector. Should be evaluated by the capital requirements, economies of scale, product differentiation, access to distribution channels, implementation costs, know-how, legal/regulatory barriers, and switching costs. For those already on the market, it is important that the entry of new competitors is difficult, rebutting their market share only against competitors already established.

The **bargaining power of buyers dimension** reflects the extent to which consumer behavior may or may not influence the company. In this sense, it should take into account the turnover per customer and the buyer's price sensitivity, which can create some dependency. When there are substitutes or a great competition via price, dependence on few customers is also a dangerous situation. On the other side, if there is a greater dependence from customers to the company's product or service, the greater is the power of the company holding as a supplier, exposing the client to the conditions demanded by the company.

The **bargaining power of suppliers dimension** is a factor, although some controllable, can negatively influence the whole structure. Therefore, the company should assess whether there are available various raw materials, suppliers, etc. The idea here is reducing dependence and control the impact of material in the production costs and profit margins. Here too, the company should consider if it is useful to consider the commercialization of large sales volume from a supplier, in order to get lower prices and short delivery time.

The **threat of substitute products dimension** is also a factor that can dictate the degree of attractiveness of the sector, since it influences the retention of customers, turning more difficult the sales forecast. The IT sector is characterized by the potential that offers to replace traditional products and services providing faster and at lower cost.

The **rivalry among existing competitors dimension** must be analyzed. The existence of a high number of competitors may indicate a fragmented market with many undifferentiated products and major problems in customer loyalty. Conversely, if the market is composed of few companies, there may be barriers to entry, making it difficult to increase market share. In the IT sector, new competitors can emerge very quickly and operating in the global market. In this sense, the company must be prepared for a high market dynamics, not only in terms of needs but also in existing and potential competitors.

2.1.3 SWOT Analysis

SWOT analysis is a management tool widely used by companies for strategic diagnosis. It is also very useful in the context of a business plan research and preparation. The SWOT term is composed of the initials of the words strengths, weaknesses, opportunities and threats.

The SWOT analysis allows:

- Making an overview of internal and external reviews;
- Identifying the key elements of the company's management, allowing to establish priorities for action;
- Preparing strategic options. SWOT analysis allows to identify the risks that should be taken into account and the problems to be resolved, as well as the advantages and opportunities to enhance and explore;
- Being a key element in the sales forecast in conjunction with market conditions and the company's capabilities.

SWOT analysis should be made and interpreted in an integrated manner, combining the elements of internal and external analysis. Therefore, the diagnosis that results from it is reliable and constitutes a source of information and adequate support to the needs of strategic management, which deals with decisions that shape the future in the medium and long terms of the organization.

The internal environment can be controlled by the organization's leaders, since it is the result of action strategies defined by them. Thus, when we noticed a strong point in our analysis, which should be highlighted further; when we perceive a weakness, we must act to control it or at least minimize its effect.

On the other side, the external environment is totally out of control of the organization. However, that does not mean it is not useful to know him. While we cannot control it, we can monitor it to seek opportunities more quickly and efficiently, and avoid threats when possible.

The SWOT matrix (see Table 2.1) consists in assessing the competitive position of a company in the market through the use of an array of two axes, each of them offering two variations: strengths and weaknesses of internal analysis; opportunities and threats of the external review. By building the matrix variable, facilitating their analysis, and looking for suggestions for decision-making, being an essential tool in training plans and defining business strategies.

SWOT analysis should be as far as possible, dynamic, and permanent. Besides the analysis of the current situation, it is important to confront it with past situations, their evolution, the expected situation, and its future evolution. This matrix suggests the obvious choice of strategies that lead to the maximization of opportunities of the environment and built on the strengths of the company and minimizing threats and reducing the effects of the weaknesses of the company.

Table 2.1   Strategies associated with SWOT matrix

	Strengths	Weaknesses
Opportunities	**SO (maxi-maxi)**	**WO (mini-max)**
	Use the organization's strengths to take advantages of opportunities	Develop strategies to minimize the negative effects of weak points and, simultaneously, take advantage of emergent opportunities
Threats	**ST (maxi-mini)**	**WT (mini-mini)**
	Use of strengths to cope with threats or to avoid threats	The strategies to be developed should minimize or overtake the weaknesses and threats

In order to guarantee whether the SWOT analysis is correctly performed, we should guarantee that the information is recent, reputable, and free of errors. We can use some techniques such as brainstorming, focus groups, interviews, and surveys. Finally, SWOT analysis should not be looked in absolute terms (all depends on the environment). In fact, an opportunity can also be a threat, or a strength can be a weakness in another context.

Figure 2.1 provides an example of some elements that can be considered in each of the four dimensions of a SWOT analysis.

It is relevant to highlight that some elements in a SWOT analysis can be seen by the company as an opportunity or a threat. For example, the appearance of new technology can be an advantage if the company is able to bring it to the conception of its products or services provided. However, it can become a threat if it is used by our competitors to create competitive advantage against our positioning in the market.

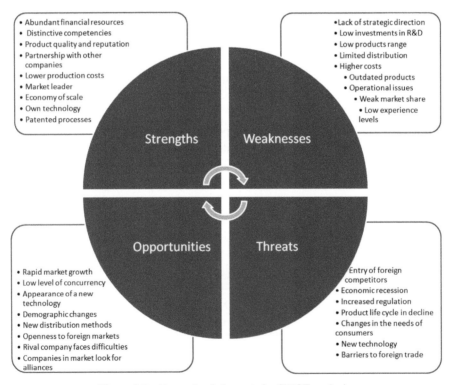

Figure 2.1   Example of elements for SWOT analysis.

2.1.4 Customers Identification

Customers' identification intends to identify who will be the customers of the company. This process is composed of three sub-sections:

1. Customer segmentation – define the target customers and how they behave;
2. Key customers – identify the key customers of the company. These can be large consumers or just individuals, whose satisfaction is key to the success of the business;
3. Customer management – define how the company will maintain a good relationship with his customers and what techniques will be used.

The **customer segmentation** allows finding groups of companies and people that the products offered by the company are more useful, which gives the power to define a specific communication language and techniques to gain the attention of customers. The result of a good segmentation potentially increases the number of sales and profits, while guaranteeing a high customers retention rate.

The use of segmentation recommends the adoption of segments based on the following principles:

- Relevant – the segment needs to be significant considering the company offer;
- Numerous – the segment is large enough (in terms of size and potential purchasers) to gain attention from the company;
- Accessible – the company has the means to achieve the segment with marketing and sales efforts;
- Profitable – the segment has the ability to pay for that demand;
- Focus – the segment gives greater objectivity of its search business, because the company builds a specific offer for each segment.

There are several types of segmentation. Among them, we highlight:

- No segmentation – the product is so segmented that the communication is unique. Examples of such situation are emails and e-books that are sent to all database contacts;
- Simple segmentation based on a single characteristic of clients;
- Segmentation based on the needs of the group;
- Segmentation attending current and potential value of the customers;
- Segmentation in different economic fields (e.g., services, industry, commerce, etc.);
- Segmentation by lifestyle (e.g., university students, people who live alone, elderly, etc.).

Besides that, the segmentation can be concentrated or differentiated. In the former approach, we design a promotional message that communicated the benefits desired by a single specific segment; in the latter approach, we design more than one promotional message, with each communicating different benefits.

The segmentation tends to be differentiated and customized in the IT field. Segmentation by lifestyle is an emerging practice and allows the company to offer a fairly focused value proposition. Thus, communication with the customer can be personalized taking advantage of the benefits and potential offered by social networks.

The identification of **key customers** is a fundamental step to offer personalized services to a specific focus group. Some customers demand more attention and care due to their relevance and importance to the company.

Analyzing customers allows the identification of those who best fit the business priorities. These key customers also depend on the company strategy. For example, a company that has recently launched a new product typically aims to identify potential customers to increase sales; on the other side, if the company has cash flow problems, the company should attract customers who pay quickly.

Typically, the customers who tend to be more profitable for the company have the following characteristics:

- Buy high-margin products;
- Pay full price without negotiating discounts;
- Place a small number of larger orders rather than many small orders;
- Do not cancel or amend orders;
- Pay on time, without being chased for payment;
- Do not require extensive after-sales service.

The **customer management** is a valuable process to understand and analyze the customers of the company. The idea is to collect data that let us identify customers and how they behave. This will let the company to find what customers think about its products and services.

The management of customer relationships and the implementation of a customer relationship management, or CRM, is a company-wide process designed to increase profitability by building customer loyalty and retention. The central point of a CRM solution is generally the creation of a single shared customer database, which allows the information to be collected once but used many times. The sharing of this customer data across the authority, in conjunction with the functional tools provided by a CRM solution, allows the authority to make gains in both efficiency and effectiveness, e.g., by improving the ability of front line staff to resolve issues at first contact or dealing automatically with inquiries that originate over the Web.

Another situation that CRM helps is capturing and analyzing customer feedback. Collecting customer feedback requires input such as systems, resources, time, and effort. From a strategic perspective, collecting customer feedback is part of the wider continuous improvement and learning process and also provides data for operational and departmental functions of the company.

CRM process brings important benefits to companies. Among them, we highlight:

- Enhanced customization of services and product offerings;
- Enable communication through multiple selling channels;
- Identify the most valuable customers and increase customer retention;
- Deliver the right product and service at the right time from the right channel;
- Channels for customers to give feedback;
- Being more responsive to customer requirements;
- Enable customer knowledge management;
- Improve product and business innovations.

2.1.5 Competitors Identification

The identification of competitors can follow two complementary approaches. The first is demand-side-based, composed of firms satisfying the same set of customer needs. The second approach is supply-side-based, identifying firms whose resource base, technology, and operations are similar to what does the local firm. Additionally, the company must pay attention to not only today's immediate competitors, but also potential competitors.

In order to analyze the competitors we must look into the following areas:

- What is their turnover – know the turnover of competitors will be important to assess what are their capabilities and size;
- What is their market share – competitors can have different or identical market shares. It is particularly important to look more in detail for competitors with identical market shares, because in principle, will be with them that your business will compete. Knowing this information can be difficult;
- What is their experience in the market – knowing the history and reputation of competitors is also relevant to know the market behavior

in the field. It is also relevant to know if competitors have been target of concentration movements in the market;

- What is their annual growth – look how has been the development of competition, that is, if they grow more than the market or other competitors;
- What are their distribution methods – analyze which are the distribution channels that competition uses, if they focus more on any in particular or not. Anyway, it is useful to explore all distribution channels;
- What are their communication techniques – analyze how they communicate with their customers, namely which media supports their use and if they make institutional advertising;
- What is the structure of their sales department – look how the sales of the competitors are made, namely if they use the Internet, sell by catalog, or they have a sales team;
- What is their pricing policy – analyze the established price of their products or services. Look also if they fight for low-cost products/services or adopt a differentiation policy;
- What is their product policy – look if the competitors typically launch regularly new products in the markets, if they have all product ranges, or if they are dependent on a single product.

After answering these questions, we will have a clear view of the competitors in the market. Furthermore, it is possible to do a SWOT analysis of each competitor.

Finally, it may also be important to identify potential competitors. The process should start by identifying firms for whom the various barriers to entry to the industry are low or easily surmountable. These may include the following:

- Technology – firms which possess the technology necessary to operate in an industry represent one source of potential competitors;
- Market access – in businesses where market access is a key factor for success, firms that operate in a different field but have access to the same market, and can become a potential competitor;
- Reputation and image – brand extension strategies are based on the use of a firm's reputation in one product area to leverage its entry into another. Typically these companies will have the potential to attract clients more easily than new emergent companies that have recently appeared in the market;

- Operating knowledge and skills – it typically involves regional competitors in a business that have limited market shares. However, based on their experience and operational excellence, they can attract to expand geographically.

2.1.6 Partners Identification

In an increasingly competitive market where consumers are highly demanding, the establishment of partnerships becomes an important ally for companies that want to increase their value and competitive edge. Business partnerships are suitable for not only big companies, but also micro, small, and medium-sized companies, offering numerous benefits for both.

Partnership can be defined as the cooperation between different organizations in order to provide mutual benefits between the involved parties. It should be a "win-to-win" relationship, in which the partner companies aim the development of a joint project that will have a positive impact on both companies. In addition, through partnerships, it is possible to optimize different processes in both companies, without losing shareholder control.

Partnerships enable companies to succeed without major investments in different fields, such as: (i) entering in different markets (including international markets), (ii) increasing market share, or (iii) improving processes and technologies in the organization. It should be noted that there is not an ideal type, or a correct model, of partnership. Instead, what exists are appropriate alliances (or not) to the needs of each company and its goals. Like any new project, partnerships also need to be well formulated and planned. Thus, it is very important to conduct a research work to assess the viability of the alliance, to guarantee that it is positive for both parties.

The research study about partnerships shall look to six dimensions:

- Established date – collect information regarding when the company was established;
- Size – number of staff and/or turnover;
- Market share – estimated percentage of market share;
- Benefits for your company – explain the unique value to your company (e.g., quality, price, or service);
- Benefits for your partner – explain the unique value offered by your company to your partner company (e.g., quality, price, or service);
- Potential issues – identify the main potential issues in a relationship between your company and each potential partner. Propose also a mitigate plan in order to respond to those potential issues.

A partnership is an agreement to do something together that will benefit all involved, bringing results that could not be achieved by a single partner operating alone and reducing duplication of efforts. Therefore, and in order to be efficient, a partnership should have a recognizable and autonomous structure to help establish its identity. It should offer several degrees of autonomy and flexibility, while preserving the lines of communication to ensure that all partners are kept informed and involved. Additionally, partnerships need to develop a long-term strategy if they are working effectively and have a lasting effect.

Some characteristics of well-established partnerships can be identified. We highlight the following elements:

- Common objectives and targets are properly set and clearly defined;
- Responsibilities and the nature of co-operation are clarified;
- Agreements are based on identifiable responsibilities, joint rights, and obligations, and are signed by all relevant partners;
- There is a sharing of risk, responsibility, accountability, or benefits;
- Partnership members have appropriate training to identify issues or resolve internal conflicts;
- The partnership takes an inclusive approach (relevant actors are involved in planning and implementation);
- A "learning culture" is fostered, i.e., one where all partners are able to learn from one another by allowing new ideas to come forward in an open exchange of experiences;
- Measures for permanent monitoring and evaluation are planned.

A good partnership in the IT field goes beyond simple advice, and needs detection and solution proposals. IT partner companies should take joint responsibility for the implementation or redirection of strategies, systems integration, and the development and availability of practical tools, in other actions, which solve observed problems and evolve desired scenarios.

For small ventures outside the IT field, it becomes impracticable to maintain professionals or an area of information technology. IT teams tend to be expensive. In addition to infrastructure issues, such as space, equipment, and software, it takes a lot of time to tailor and train professionals according to the area of activity and business objectives. The use of partnerships in information technology allows reducing risks and costs. With much less costly investment, the role of partnerships in the IT field is to ensure the support the company needs, together with co-responsibility for the achievement of defined objectives.

2.1.7 Strategic Positioning

The definition of a strategic direction aims to coordinate the business activities and optimize the resources used, in a way that customers observe and interpret this strategy in relative terms, comparing it to the company's competitors.

The term "positioning" is used with different approaches. However, there is a consensus that one of the most important elements is the perception of the company positioning by their current and potential customers. In fact, this term refers to the image obtained by a product in the consumer's mind, as a result of three dimensions: (a) the type of offer that the company does; (b) the target audience of the offer; and (c) competition.

The strategic positioning of the company needs the execution of two previous actions: (a) market segmentation and (b) identification of the target market. Organizing it, in order to perform a strategic positioning, we need to perform the following six steps:

1. Identification of the segment variables and market segmentation;
2. Development of profiles/characteristics of the resulting segments;
3. Assessment of the attractiveness of each segment;
4. Selection of the market(s) segment(s);
5. Identification of possible positioning concepts for each target market;
6. Selection, development, and communication of the chosen positioning concepts.

The company should be particularly interested in choosing only some difference in the target market, in order to receive consumer preference at the time of purchase. There are different product positioning strategies in their market segment, namely:

- Positioning by attribute – occurs when the company establishes its positioning based on the performance of their products in some specific attributes (tangible or abstract);
- Positioning by benefits – not only looks to the product performance in certain attributes, but also highlights the benefits they bring;
- Positioning by use – presents the product as more suitable to be used or applied in a specific situation;
- Positioning by user – associates the product with a specific user profile, based on characteristics such as lifestyle, personality traits, life history, etc.;

- Positioning by competitor – compare explicitly or implicitly the products with other competitors, seeking to facilitate the perception of customers of the company positioning. The idea is that customers perceive that the company products are a better alternative than the competitor;
- Positioning by product category – the product is positioned as a leader in a particular product category. It is used when it comes to new products, allowing to highlight the differences in characteristics between them and the other products on the market.

A great tool to visually understand the positioning of a company among his competitors is the adoption of a strategy canvas. The purpose of the strategy canvas is to help identify blue ocean opportunities that the company can dominate. This includes identifying those untapped markets. This tool has the advantage to capture the current state of play in the existing market, allowing to realize the factors on which the industry competes and where competitors are investing. Furthermore, this tool allows comparing the areas where the company diverges from the competition in order to open new tangential markets. An example of a generic strategy canvas is given in Figure 2.2.

Porter has described a category scheme consisting of three general types of strategies that are commonly used by businesses to achieve and maintain competitive advantage. These three generic strategies are defined

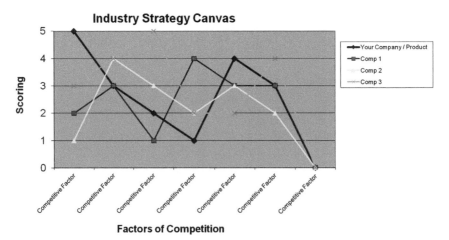

Figure 2.2   Strategic canvas example.

along two dimensions: strategic scope and strategic strength. Strategic scope is a demand-side dimension and looks at the size and composition of the target market; on the other side, strategic strength is a supply-side dimension and looks at the strength or core competency of the firm. Generic strategies adopted to overcome the competitors are divided into three categories:

- Cost leadership – it is based on achieving lower costs than competition, using efficiency as a key factor;
- Differentiation – it is based on the achievement of higher-value products comparing to the competition, because different attributes as quality, innovation, intimacy with customers, etc. are used in order to create distinctive key factors;
- Focus or specialization – it is based on the selection of a limited segment of customers, which the company can serve better than the competitors, in differentiation or efficiency. It is a typical strategy adopted for small businesses.

The adoption of a generic strategy by the company affects its organization. An overview of the resources, aptitudes, organizational requirements, and risks associated with each generic Porter strategy are presented in Table 2.2.

Table 2.2 Structure of the company based on the generic Porter strategy

	Resources and Aptitudes	Organizational Requirements	Risks or Limitations
Cost Leadership	- Sustained investment of capital and favorable access to financial markets - Special aptitudes for process engineering with close supervision - Products designed for ease of manufacturing - Low cost of distribution	- Strict control of costs - Detailed and frequent control reports - Clearly defined organization and responsibilities - Incentives based on meeting quantitative objectives	- Technological change that cancels out the experience gained or investment made - Competitors who learn easily and rapidly - Stagnation of the product or market - Inflation of costs that annuls the previous price differential

(Continued)

Table 2.2   Continued

	Resources and Aptitudes	Organizational Requirements	Risks or Limitations
Differentiation	- Significant aptitudes in marketing and in product engineering - Strong investment in R&D - Prestige in quality and technology - Full cooperation of the distribution channels - Long tradition in the sector or a unique combination of distinctive aptitudes	- Coordination between the functions of R&D, product development, and marketing - Qualitative assessments and incentives - Capacity for understanding the market and how it changes - Stimulus for creativity	- Different competitive levels of product prices - The customers no longer value the product's factors of differentiation - As the industry matures, imitation reduces the perceived differentiation
Focus or specialization	- Resources and aptitudes in a special field of the company's area of operation - Dominance of the relevant technology and of the engineering of the product - Marketing capacity - Ability in the use of limited resources	- Flexible and efficient organization structure - Corporate culture relevant and specific to its areas of specialization (products and markets) - Close coordination between functions - Rapid response to changes in the environment	- The differences in costs compared with non-specialized companies are so wide that the advantages of specialization are eliminated - The market in which the company is specialized reduces its differences with respect to the global market - Other competitors are specialized in part of the market of the already specialized company

A good example of a company that adopts a cost leadership strategy is Walmart or Tesco; in differentiation strategy, we find companies such as FedEx or Apple; finally, good examples of companies adopting a focus strategy are Mixpanel or Quidsi.

For a small venture operating in the IT field, a typical strategy is to adopt a focus approach. The use of information technologies allows improving existing processes, making them more agile and offering better results. It is also important to recognize the role of social networks in the adoption of a focus approach, allowing the message to be customized according to the target audience. In addition to this customization, social networks also allow real-time tracking of a campaign's success, allowing the company to make the necessary corrections in a timely manner.

2.1.8 Advertising & Sales

Advertising may be defined as the process of buying sponsor-identified media space or time in order to promote a product or an idea. The idea behind advertising is to persuade the buyer to buy products with a view to maximizing a company's profits. Advertising is composed by the following elements:

- It is a mass communication reaching a potential large group of consumers;
- The communication message may be personalized or not;
- The communication is speedy, allowing a rapid diffusion of the message and it is economical, which keeps a low cost per message.

Advertising is a form of promotion that should be specifically attending the target market, in order to be effective, not just for one customer, but for many target buyers. This requires that the target consumers should be specifically identified and the effect which advertising is intended to have upon the consumer should be clearly indicated.

The basic objectives of an advertising program include:

- Stimulate sales amongst present, former, and future consumers;
- Communication with consumers;
- Retain the loyalty of present and former consumers;
- Increase support by bolsters the morale of sales force, distributors, wholesalers, and retailers;
- Project an image of respect and trust for an organization.

There are different types of advertising mediums available in the market with their inherent advantages and disadvantages. When choosing the advertising mediums to use, companies must understand who their target audience is and which most effective methods of reaching them are available. A good option typically consists in dividing advertising budgets among the various media resources in order to reach the most customers with different target groups.

When deciding an advertising medium, several factors should be considered:
(i) reach of the media; (ii) nature of the product; (iii) position in product life
cycle; (iv) cost of each media channel; (v) size of advertising budget; and
(vi) online or offline mode.

Table 2.3 summarizes the main advantages and disadvantages of each
advertising channel.

Table 2.3   Characteristics of each advertising channel

	Target Audience	Advantages	Disadvantages
Television	- General public	- Wide reach - Audio and visual messages - Good for simple messages and slogans	- Expensive - Limited length of exposure - Absence of interaction
Radio	- General public	- Medium to high reach - Good for simple messages and slogans - Less expensive than TV	- Absence of interaction - Still expensive compared with Internet mediums - Limited length of exposure - Audio messages only
Newspaper	- General public (but only to literate segment)	- Wide reach - The message can be reviewed and re-read as needed	- Less audience compared to TV or radio - Ad space can be expensive - Publication depends on the whim of the editors
Magazine	- Specific public	- The message can be reviewed and re-read as needed - High reader involvement - Detailed information	- Long lead times - Limited flexibility - Space and ad layout costs are higher
Internet/ websites	- General public	- No limitation for local or regional zones - Less expensive - Detailed information	- Not easy to maintain updated
Social media	- General public and specific group	- Great for building customer relationship - The message can be easily customized - High potential for viral marketing - Less expensive	- Audience targeting is low - No control of the message diffusion

Table 2.3 Continued

	Target Audience	Advantages	Disadvantages
Mobile phones	- General public (but teenagers segment is more attractive)	- Low cost - Highly popular	- The message should be short
Brochures & flyers	- General public and specific group	- Suitable for instructional info - Less expensive - Can be mailed - Can be produced in-house	- Limited to specific distributions
Newsletters	- General public and specific group	- Can deliver more information than brochures and flyers - Can be produced in-house	- Limited to specific distributions
Public presentations	- Specific public	- Encourages group formation - High level of interaction - Builds partnership	- Need to be held when people are available - Difficult to reach a wide target audience - Do not always attract desired audience
Telemarketing	- Specific public	- High level of interaction - The results are highly measurable - Can be easily outsourced	- Aversion to telemarketing - High costs - More suitable to contact current customers than make new customers
Promotional items (e.g., t-shirts, shopping bags, cups, etc.)	- General public	- High flexibility and creativity - Moderately inexpensive due to economies of scale - Popular among teenagers	- Can be costly to produce - Difficult to identify the target audience - High logistic costs

Another point that should be presented in the business plan is the identification of the distribution channels used and their impact on the sales. The functional aspect of the distribution channel is seen as a way of connecting and ordering of agencies and intermediaries through which one or more streams are flowing. The most important streams in distribution channels are: (i) physical movement of completed products or services; (ii) actual transfer of ownership laws among participants of the channel; (iii) information about potential buyers, competition and demand; (iv) promotion; (v) payments of invoices; (vi) negotiations; (vii) realization of orders; (viii) risk taking; and (ix) shipping, transportation, and storage of goods.

The decision of distribution channel includes two structural systems: vertical and horizontal. In the vertical structure, we can specify the number of different levels of flow streams; in the horizontal structure, we can determine the number and type of intermediaries on specific levels.

In general, the company can choose between direct and indirect channels. In a direct channel, the company sells directly the goods to consumer without the use of intermediaries. On the other hand, the indirect channel is characterized by the existence of intermediaries, where several-level channels (one level, two levels, three levels, etc.) can exist. In these levels, models appear entities such as wholesalers, retailers, and agents.

Typically, business products tend to move through shorter channels than consumer products due to geographical concentrations and comparatively few business purchases. Service firms market primarily through short channels because they sell intangible products and need to maintain personal relationships within their channels. Table 2.4 presents the main factors that influence the marketing channel strategies.

Table 2.4   Factors influencing marketing channel strategies

	Characteristics of Short Channels	Characteristics of Long Channels
Market Factors	- Big business users with large orders - Geographically concentrated - Extensive technical knowledge and regular servicing required	- Consumers with small orders - Geographically dispersed - Little technical knowledge and regular servicing not required
Product factors	- Perishable - Complex - Expensive	- Durable - Standardized - Inexpensive
Producer factors	- Manufacturer has adequate resource to perform channel functions - Broad product line - Channel control by providers is important	- Manufacturer lacks adequate resources and knowledge to perform channel functions - Limited product lines - Channel control is not important

2.1.9 Marketing Mix

The concept of marketing mix (or 4P's) condenses four variables: (i) product, (ii) price, (iii) distribution or place, and (iv) promotion. Together, these four components constitute the marketing mix of a company that together enables the provision of a product with the characteristics desired by consumers.

The **product policy** refers to the characteristics and attributes of the products and services offered by the company. Features in general should be defined, such as technical features, quality, brand, packing, size, and colors, among others.

The **price policy** refers to how much and how will be charged to the customer. In this regard, the product can be cheap or luxury. It can be charged at one time or monthly. Among many other strategies are psychological price, pay what you want, auctions, etc.

Table 2.5 presents different price strategies that can be adopted by companies.

Table 2.5 Price strategies

	Description
Cost-plus	Adds a standard percentage of profit above the cost of producing a product. Accurately assessing fixed and variable costs is an important part of this pricing method.
Value-based	Based on the buyer's perception of value (rather than on the products' costs). The buyer's perception depends on all aspects of the product, including non-price factors such as quality, healthfulness, and prestige.
Competitive	Based on prices charged by competing firms for competing products. It is mandatory to know very well the price strategy followed by competitors and react to price changes.
Going-rate	A price charged that is the common or going-rate in the marketplace. Going-rate pricing is common in markets where most firms have little or no control over the market price.
Skimming	Involves the introduction of a product at a high price for affluent consumers. Later, the price is decreased as the market becomes saturated.
Discount	Based on a reduction in the advertised price. A coupon is an example of a discounted price.
Loss-leader	Based on selling at a price lower than the cost of production to attract customers to the store to buy other products.
Psychological	Based on a price that looks better, for example, 9.99 € instead of 10.00 €.

After the decision on the pricing strategy, there are other aspects that should be taken in account. These elements have an impact on the construction of the price and will decrease or increase the amount of money received by the company, namely:

- Payment period – length of time before payment is received;
- Allowance – price reductions given when a retailer agrees to undertake some promotional activity;
- Seasonal allowances – reductions given when an order is placed during seasons that typically have low sales volumes to entice customers to buy during slow times;
- Bundling of products/services – offering an array of products together;
- Trade discounts – payments to distribution channel members for performing some function such as warehousing and shelf stocking;
- Price flexibility – ability of the salesperson or reseller to modify pricing, based on different customer groups or geographic regions;
- Volume discounts and wholesale pricing – price reductions given for large purchases;
- Early payment discounts – policies to speed payment and thereby provide liquidity;
- Credit terms – policies that allow customers to pay for products at a later date.

The **distribution** or **place policy** refers to the channels used to make the product available to the customers. The distribution takes place where we sell the product, distributors and carriers to use, and storage.

The **promotion policy** refers to the dissemination strategy in terms of communication. Therefore, it is responsible to make the products and services visible to the customers. For instance, it can be used in different dissemination strategies such as TV spots, radio, Internet, etc.

Table 2.6 provides a brief overview of the questions that should be answered in each component of the marketing mix.

The standard concept of 4Ps has evolved by several authors, which introduces three new variables, thereby creating the 7Ps marketing model. This model proves to be more useful in the service sector and in environments saturated with information.

The 7PS marketing model is, therefore, composed of three additional components: (i) personnel or people, (ii) processes, and (iii) physical assets.

The **personnel policy** is a main component of the services sector. Since consumer cooperatives are firms in which employees face with consumers

Table 2.6 Questions to be answered by marketing mix

	Questions
Product	- What does the customer want from the product or services?
	- What features the product should have?
	- How the customer will use it?
	- What will be its appearance?
	- How is it differentiated versus your competitors?
Price	- What is the value of the product or service to the customer?
	- Are there reference prices in the area?
	- Is the customer price sensitive?
	- What discounts should be offered to customers?
	- How will be your price compared to competitors?
Place	- Where do customers look for your product or service?
	- How can your company access the right distribution channels?
	- Does the company need to use a sales force?
	- What do your competitors do and how can you learn from them in order to have a differentiate offer?
Promotion	- How and when will be announced the company's products or services?
	- Which promotional channels will be used to reach your audience?
	- When is the best time to promote?
	- How do your competitors do their promotions?

directly, such organizations try to achieve a special situation in the market through training their employees on sales knowledge and how to treat with customers.

The **processes policy** ensures availability and sustainable quality of services. The task and role of this component of the marketing mix is to balance service demand and supply. By improving the procedure of providing services to customers, cooperatives can pave the ground for consumers' convenience, which leads into repurchase and, finally, sales increase.

The **physical assets policy** refers to the environment and facilities needed by companies to provide services to their customers. Consumer cooperatives can expand consumers' choices by providing facilities like self-service, paramount shelves, etc. same as big shops and can prevent a buyer leaving the company without any purchase.

Several authors also consider introducing 8Ps in the marketing mix. The final P corresponds to the **productivity & quality policy**. It refers to the capacity of the company to offer good deals to the customer. Therefore, it also refers to the capacity of the company to provide and maintain a good image and reputation with their customers.

2.1.10 Business Model Canvas

A very useful tool that can be used to get an overview of the business structure is the business model canvas. It is a visual template composed of nine blocks of a business model (Figure 2.3). Together these elements provide a pretty coherent view of a business' key drivers. The blocks are:

- Customer segments: Who are the customers? What do they think, want, or do?
- Value propositions: What is compelling about the proposition? Why do customers buy or use?
- Channels: How are these propositions promoted, sold, and delivered? Why? Is it working?
- Customer relationships: How does your business interact with the customers?
- Revenue streams: How does the business earn revenue from the value propositions?
- Key activities: What uniquely strategic things does the business do to deliver its proposition?
- Key resources: What unique strategic assets must the business have to compete?

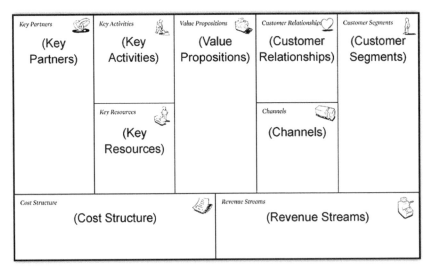

Figure 2.3   Nine elements of business model canvas proposed by Osterwalder and Pigneur.

- Key partnerships: What can the company not do so it can focus on its key activities?
- Cost structure: What are the business' major cost drivers? How are they linked to revenue?

In the IT field, all nine elements of a business model canvas must be filled like in other type of businesses. However, greater difficulties can be experienced in filling in the "key resources" block. In this block, and considering the specificities of the IT sector, not only material resources should be specified, but also the human resources that are vital for the execution of the key activities performed by the company.

2.2 Scenario I – TourMCard

2.2.1 Market Research

In order to estimate the market size and needs in the sector of travelling and tourism, we adopt simultaneously qualitative and quantitative approach. For that, we searched for public reports from consultant companies and government agencies about the dimension, characteristics, and particularities of the tourism sector. The following documents were considered:

- The Travel & Tourism report for 2015 from the World Travel & Tourism Council (WTTC);
- An analysis of travel and tourism sector related to 2013 performed by KPMG;
- The Travel & Tourism Competitiveness report for 2013 available from the World Economic Forum.

Based on a qualitative research, it was possible to extract the following conclusions:

- Travel & Tourism generated US $7.6 trillion (10% of global GDP) and 277 million jobs (1 in 11 jobs) for the global economy in 2014 (WTTC, 2015);
- The tourism sector is growing at a higher rate than other significant sectors, such as automotive, financial services, and health care (WTTC, 2015);
- Investments in the tourism sector in 2014 were US $814.4 bn. It should rise by 4.8% in 2015 and by 4.6% over the next 10 years (WTTC, 2015);
- New tourist destinations, especially those in emerging markets have started gaining prominence. Asia Pacific offers the highest growth in

the number of international tourist arrivals, followed by Africa (KPMG, 2013);

- Arrivals for the purpose of visiting friends and relatives, health, religion, etc. are expected to witness faster growth than those for business and professional purposes (KPMG, 2013);
- Looking for the Travel & Tourism Competitiveness Index (TTCI) in 2013, it is possible to extract the following conclusion for each region. Europe remains the leading region for travel & tourism competitiveness, with all of the top five places taken by European countries. The first three places are occupied by Switzerland, Germany, and Austria. Spain is the country among the top 10 that sees the most improvement since 2011. In the Americas region, the United States is the highest-ranked country. The others two places in podium are occupied by Canada and Barbados. In the Asia-Pacific region, the first three places are occupied by Singapore, Australia, and New Zealand; In the Middle East and North Africa region, the first three places are occupied by the United Arab Emirates (UAE), Qatar, and Israel. In the sub-Saharan region, the first three places are occupied by Seychelles, Mauritius, and South Africa (WEF, 2013);
- The key capabilities that drive travel & tourism stable growth performance are affinity to travel & tourism, policy rules and regulations, price competitiveness, environmental sustainability, and safety and security (WEF, 2013).

Figure 2.4 presents a distribution of the international tourist arrivals grouped by each region (Europe, Asia, Americas, and Africa).

Additionally, qualitative studies have shown that international tourist arrivals are set to increase at a growth rate of 3.3% per annum. By 2020, 1.4 billion tourists and by 2030, 1.8 billion tourists are expected (KPMG, 2013).

2.2.2 Environmental/Industry Analysis

We will start by making the PEST analysis, which is depicted in Table 2.7.

The PEST analysis includes several common items to all businesses. However, there are others that are specific to the tourism market. In **political dimension,** the government support in tourism development may enable the company to receive financial funds as well as participation in fairs and touristy promotion campaigns in international markets. In **economic dimension,** lower costs of transportation will increase the mobility of people.

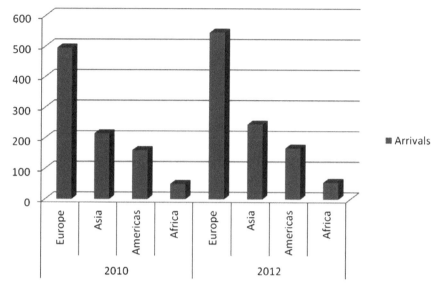

Figure 2.4   International tourist arrivals (in million).

Table 2.7   PEST analysis of TourMCard

Political	Economic
- Political instability in several European countries	- Exchange rate fluctuation
- Government support of tourism development	- Economic stagnation
- Weak intellectual property protection in emerging markets	- Rising energy costs
- High governmental pressure for carbon targets	- Low cost of transportations
- Increase importance of EU legislation	- Difficulties in credit access

Social	Technological
- High employment rate	- Easy access to media
- Increased interest in leisure	- Increase of low cost air travel
- Low illiteracy rate in developed countries	- Easier access to the Internet
- Increased senior population	- Increased importance of eCommerce
- Declining birth rates in developed countries	- Increasing importance of open-source technologies
- Different lifestyles	- Adoption of open standards
	- Eco-concerns on product packaging materials
	- Mobile and cloud-based services are gaining more importance

Additionally, it will allow people to spend more money on other services related to tourism. In **social dimension,** the existence of different life cycles causes the appearance of different types of tourism, such as rural tourism, nature tourism, adventurous tourism, etc. Finally, in **technological dimension**, the emergence of low-cost air travels caused the increase in travel, in the number of visits, but also turned possible for people with lower financial resources to travel.

Then, we will show the Porter's Five Forces, which is depicted in Table 2.8.

The **threat of new entrants** is **high** due to high industry attractiveness and no regulation in the market; the **bargaining power of buyers** is **moderate** because buyers are very sensitive to the price and there are several competitors in the market, even if the number of buyers is typically higher compared to the number of sellers; the **bargaining power of suppliers** is **low** because there are a lot of different suppliers, including the open-source market; the **threat of substitute products** is **high** due to the emergence of new technologies based on cloud and mobile services that can, in short time, make your service obsolete; finally, **the rivalry among existing competitors** is **high** because there are a high diversity of segmented competitors, even if the customer loyalty could be a barrier to attract incumbent customers.

Table 2.8   Porter's Five Forces analysis of TourMCard

Threat of new entrants	- Low protection mechanisms - Easy to offer product differentiation - Easy to enter a local player - Industry growth rate is high - Need of specialized knowledge in IT
Bargaining power of buyers	- Number of buyers are much greater than sellers - Switching costs are low - Price sensitivity is high
Bargaining power of suppliers	- Many alternatives for suppliers - High availability of substitute technologies - Costs of changing are low
Threat of substitute products	- Emergence of cloud services - Emergence of mobile services
Rivalry among existing competitors	- Fixed costs are very low - High diversity of competitors - High segmentation - Customer loyalty

2.2.3 SWOT Analysis

The SWOT analysis is depicted in Table 2.9. The strengths are essentially in three levels: low costs, human resources, and functionalities of the application. On the other hand, the weaknesses are essentially common to the majority of startup companies. The opportunities appear in four dimensions: market, distribution channels, technologies, and partners. On the other hand, threats appear in different fields. It is also important to realize that partners and technologies can be seen as having a positive or negative impact on the company. In fact, the appearance of new technologies gives an opportunity for our company to create high tech solutions, but on the other hand, they can be exploited by competitors. In the same way, partners give us the possibility to offer a more complete and attractive solution for our customers, but at the same time, in medium and long terms, we will start by seeing a high dependence on them and our business could be affected by an eventual negative performance of our partners.

Table 2.9 SWOT analysis of TourMCard

Strengths	- Low fixed costs
	- Low distribution costs
	- Qualified human resources and high innovation capacity
	- A flat organizational structure
	- Ability to personalize the service to different customer segments
	- High availability of services (24/7)
	- Use of open-source technology
	- Application available in several languages
Weaknesses	- Lack of experience in the field
	- Lack of brand image
	- Lack of experience of the management team
Opportunities	- Rapid market growth
	- Access to international markets
	- Appearance of new distribution channels
	- Appearance of new technologies
	- Synergy with associated partners
Threats	- Appearance of new technologies
	- Market instability
	- High dependence of partners and negative performance of partners
	- Low degree of loyalty
	- High competition level
	- Incompatibility of equipments
	- Hacking of personal information and denial of service

2.2.4 Customers Identification

The segmentation in the tourism field is very important due to its large size, which turns difficult to reach all segments efficiently. Besides that, the travel market is too diverse to communicate. Therefore, breaking up the market will make it easier to manage.

There are several alternatives to segment the tourism market, such as demographics, location of residence, value spent, technological equipment, or lifestyle. We will perform a market segmentation based on age, lifestyle, and financial status:

- Ages between 18 and 45, which represents a typical segment engaged with new emerging technological solutions;
- Type of leisure activities performed by tourists, in particular, gastronomy, landscape, culture, cruises, fun nights, etc.;
- Total expenditure by the tourists to focus only on the most profitable customers.

Our customer keys will be the tourists with age between 18 and 45 that are willing to visit cultural places and other leisure events in the city of Porto. This segmentation will let us focus our offer and customize our solution to attend this specific segment. Additionally, this segmentation will help us to identify the most adequate partners for the initial phase of our company.

2.2.5 Competitors Identification

In order to identify and analyze the competition, we will start by identifying the competitors (Table 2.10) and give more detail about them after that.

Then, we provide detailed information about each competitor.

Tourism of Porto

The City of Porto and the entire northern region of Portugal are internationally famous for huge attractions that combine tradition with modernity in a way that few European cities can match. For these reasons, the Tourism of Porto institution created the Porto Card.

The Porto Card is a pass created to help tourists to discover the best of the Porto City and can be purchased from 6 EUR. The tourists may have access to various benefits such as free tickets in 18 museums, including Serralves, and guided tours in Music House and Stock Exchange Palace, discounts between 15% and 20% in restaurants, and 25% discounts on trips through the panoramic bus. There are also discounts on cruises, wine caves, shopping, restaurants, nightclubs, and bar shows.

Table 2.10   Competition overview of TourMCard

Competitor	Activity Start Date	Market Share (%)	Strong Points	Weak Points
Tourism of Porto	2014	Not available	- Access limited to some museums and restaurants - Free entrance and discounts	- Limited availability of services - Low level of partners - Online purchases unavailable
Tourism of Lisbon	1997	Not available	- Free entrance in some museums and others monuments - Discounts between 10% and 50% in places and cultural services - Free use of metro, bus, trams, and elevators	- Low level of partners - High price
Tourism of Algarve	2013	Not available	- Existence of several packs (nature park, transport pack, cultural pack, etc.) - Unlimited access to the Internet for 30 days	- High price - Very reduced discounts offers - Spanish language is not available in the website
Tourism of Barcelona	Not available	Not available	- Free entrance in the most relevant museums in Barcelona - No need to wait in queues (easy accessibility) - City guide available in 6 languages	- Restriction of a minimum of 3 days and maximum of 5 days per card
Bali Tourist Card	Not available	Not available	- Different types of cards for students and alumni community - Discounts vary from 15% to 30% - Existence of a forum area	- Map of the city is not detailed - No media information - Information on the site is very residual

(Continued)

Table 2.10   Continued

Competitor	Activity Start Date	Market Share (%)	Strong Points	Weak Points
Octopus Card	Not available	Not available	- Rechargeable and contactless "smart card" - Discounts vary from 5% to 10% - Variety of octopus cards - Availability at Hong Kong airport - Card design with local iconic landmarks	- It is not focused in the tourist sector - Limited access to tourist events and attractions

Tourism of Lisbon

The Lisbon Card reflects the wish of several entities to provide for the tourists of Lisbon an easy and advantageous way to know the city. It is a joint initiative from Tourism of Lisbon, IGESPAR – Institute of Asset Management Architectural and Archeological, BMI – Institute of Museums and Conservation, Carris – Trams and Buses Company of Lisbon, and CP – Portuguese Railways, also with the collaboration of other entities.

This card grants to the holder the following benefits:

- Free movement in metro lines and buses, trams, lifts, and rails in the Sintra–Rossio line and Cascais–Cais do Sodre line;
- Free admission at 25 museums, monuments, and other places of interest;
- 10%–50% discount on local tourist and cultural interests;
- 5%–10% discounts in some shops of genuinely Portuguese souvenirs;
- Available for immediate delivery;
- Sintra National Palace and Queluz with a discount of 10% and 15%, respectively, under the tariff.

Tourism of Algarve

Algarve pass is the official tourist card that allows experiencing the best that the Algarve region has to offer to its visitors. It offers a single card that represents a simple and economical way to explore the fantastic region of Algarve and allow visitors to enjoy automatic discounts at the best restaurants, tourist attractions, and many others, offering access to the main events of the Algarve. It also provides access to interurban transports or bus tours, entrances in the best museums, and even unlimited Internet access for 30 days.

The available service also allows the user to manage all movements associated with his card through his client area. The application can be accessed by Web browser from desktop interfaces or smartphones.

The concept, supported by a strong technological approach, takes over as the main premise an active contribution to the consolidation and strengthening of the brand Algarve and, consequently, to the economic growth and the attraction of tourists.

Tourism of Barcelona

The Barcelona Card is a card 3-in-1 card: ticket, museum pass, and discount card. The Barcelona Card is the official tourist pass in Barcelona. It is easy to buy the card and it helps travelers to save time and money during their stay in the city of Barcelona. Holders of Barcelona Card enjoy unlimited travel on public transports, free admission to the best museums in Barcelona, and more than 70 offers and discounts on tours, circuits, entertainment, shopping, dining, and nightlife.

This card not only allows entering in great attractions of Barcelona, but also helps to move ahead in queues for many tourist attractions, saving valuable time on short trips. Additionally, the card offers access to a city guide, available in six languages, which presents main offers (Gaudi, Sagrada Família, Camp Nou, etc.).

In full, the card offers over € 320 in savings for travelers who use the card intensively.

Octopus Card

The Octopus card is a very useful and convenient type of Hong Kong travel pass. The card is rechargeable and contactless that can be used on most forms of public transports (bus, minibus, ferry, trams, and MTR trains), as well as setting payments at all major convenience stores, such as fast food restaurants, supermarkets, bakeries, self-service vending machines, personal care stores, and major photo service outlets.

Octopus fares are sometimes between 5% and 10% cheaper than ordinary fares on most trips made on the MTR network, while reductions could apply for transits of specific bus services. It is very easy to use and the payment is deducted instantly when the user passes the card over the Octopus reader. The user can add money to the smart card at MTR stations and at most of the retail outlets that accept Octopus payment.

There are various types of Octopus cards available. The Octopus has some Hong Kong iconic landmarks featured on the card face and it can be easily obtained from the major convenience stores within the airport or in the city at a cost of HK$39. The user does not have to pay any deposit and can bring the card as souvenir afterwards and use it again when travelling to Hong Kong next time.

2.2.6 Partners Identification

There are two kinds of strategic partners for our business. In the first group appear local authorities of our destinations; in the second group emerge supply companies that produce chip/magnetic cards. A survey of some strategic partners can be visualized in Table 2.11.

2.2.7 Strategic Positioning

The *TourMCard* intends to keep a constant reference customer focus, seeking to meet the need of their customers. The company adopts a concentrated segmentation strategy by designing a common transversal message to the touristy sector. Our intention is to create an appealing and dynamic image, with customer support mechanisms, partnerships, advertising, usability, versatility, and logistic in order to differentiate from our competitors.

When comparing *TourMCard* with our main competitors, we realize that our strategic focus will be on a strong Web presence, establishment of partnerships with different cities, high usability of our solutions, and very high versatility of available services. On the other hand, we realize that our positioning is not so good in terms of price, because our price will be higher than some competitors. Advertising will be difficult to us due to financial constraints, and we will expect to get some logistic issues with our suppliers if we need a significant amount of smart cards in short term. An illustration of the above scenario is presented in Figure 2.5.

TourMCard plans to have a higher market share than our competitors. We will have a presence in several worldwide cities and accommodate a greater number of partners, which allow distinguishing us from our competitors, particularly in terms of:

- Availability of a 24 hours support service. Our order system will be accessible by Internet and/or telephone, which facilitate requests and convenience;
- Access to a wide variety of discounts;
- Wide range of partners that can thus let our company to satisfy a great number of customers.

Table 2.11 Partners overview of TourMCard

Partner	Activity Start Date	Size	Market Share (%)	Benefits for Our Company	Benefits for Partners	Potential Issues
City of Porto	Not available	23,0000 inhabitants	Not available	- Brand image - Number of potential customers - Access to touristy services - Access to public transports - Better and integrated touristy offer	- Better offer to tourists - Increase the number of potential tourists	- Duplication of offers between our service and Porto card - Investments already made on Porto card - Government bureaucracy and political instability
City of Hong Kong	Not available	Not available	Not available	- Brand image - Number of potential customers - Access to touristy services - Access to public transports - Better and integrated touristy offer	- Better offer to tourists - Increase the number of potential tourists	- Potential conflicts with Octopus Card - Lack of knowledge of local market

(Continued)

Table 2.11   Continued

Partner	Activity Start Date	Size	Market Share (%)	Benefits for Our Company	Benefits for Partners	Potential Issues
CardLogix	1998	Presence in 36 countries	Not available	- Production of seamless smart cards - Personalization of the card - Low production costs	- Increased demand of seamless smart cards	- Adaptation of supply and demand
Smart Card World	1991	Not available	Not available	- Production of seamless smart cards - Very low production costs for high quantity of cards	- Increased demand of seamless smart cards	- Difficulties in the personalization of cards - Minimum order quantity of printed cards is 1000 cards - Does not provide an integrated ecommerce solution

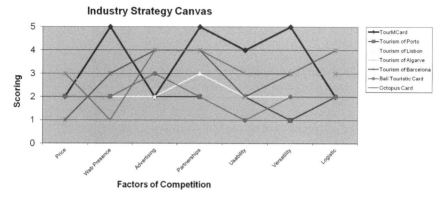

Figure 2.5   Industry Strategy Canvas of TourMCard.

Our company will adopt a competitive differentiation strategy. We will work in all touristy market, adopting a differentiation strategy in the following domains:

- Creativity of our human resources;
- Customer support service;
- High diversity and quality of local partners;
- Strong Web presence.

2.2.8 Advertising & Sales

We plan to use different kinds of advertising channels to promote our services. Table 2.12 presents the considered advertising channels.

In terms of distribution channels, we will use mainly the website to receive request order for tourists' cards. Moreover, the website is also the main element of the communication with customers, and will enable travelers to check the various available services.

Table 2.13 gives an overview about the adopted distribution channels.

2.2.9 Marketing Mix

Product/Service Policy

In terms of product policy, we offer a wide range of services and tourist packs with discounts in areas such as entertainment, culture, and leisure. Thus, it is expected that the business is promising and sustainable to guarantee quality and modern services, which are key factors for its success and implementation.

Table 2.12   Advertising channels

Advertising Channel	Benefits	Costs
Social Networks	- Increasing marketing force - Increased knowledge of the company - Better interaction with current and potential customers	Not available
E-mail marketing	- Intensification of the company advertising - More direct marketing and connected to the characteristics of the customer - Personalization of the message to specific customers - Knowledge of updates, promotions, and other benefits for customers	Not available
Partnerships	- Strengthening confidence in our brand - Increase connection between our company and partners	Not available
Tourism agencies	- Brand awareness	Not available
Advertising on buses in main airports	- Divulgation of online platform and promotional packs - Targeted marketing to tourists that come to the town	745 €/month
AdBike – Advertising on bicycles	- Divulgation of online platform and promotional packs in the main sights - Direct reference to adherence to packs, according to each touristy area	750 € (2 days/5 hours per day)
Participation in international fairs	- Divulgation of the business in Asian markets	Not available

Table 2.13   Distribution channels

Distribution Channel	Product/ Service	Percentage of Sells	Advantages	Disadvantages
Website	- Tourist card order - Organization of touristy routes - Check places of interest - Access to discounts	60%	- Easy to use the services - Availability 24/7 hours/days	- Customers need to have Internet access
Agent	- Reserve of packs	10%	- Turn the packs more attractive - Direct relationship with tourists	- Extra cost

Table 2.13   Continued

Distribution Channel	Product/ Service	Percentage of Sells	Advantages	Disadvantages
Touristy operators	- Reserve of packs	20%	- Turn the packs more attractive - Direct relationship with tourists	- Extra cost
Hotel facilities	- Reserve packs - Realizing touristy routes	10%	- Establishment of partnerships - Direct relationship with tourists	- Extra cost - Logistical difficulties to manage all events

TourMCard offers the following packs:

- Cultural pack – access to religious museums, discoveries museum, museum of electric car, among other museums and monuments;
- Night time pack – establishment of partnerships with bars, clubs, restaurants, and cafes;
- Caves pack (exclusive in Porto) – establishment of partnerships with Croft caves, Sandeman caves, Real caves, Companhia Velha caves, and Taylors cave;
- Cruise pack (exclusive in Porto and Lisbon) – partnerships with Douro Azul, ViaDouro Cruises, and Costa Cruises;
- Gift voucher – it is the ideal gift that gives total freedom to choose the perfect experience;
- Gourmet pack – memorable gourmet experiences in incomparable spaces;
- Sea & Tejo pack (exclusive in Lisbon) – perfect to make cruises in Tejo river and diving on the coast of Lisbon.

Price Policy

The main objective in terms of price policy will be competing with the remaining competitors in the market through distinctive services and appealing price. We propose the adoption of three segmented discounts:

- 5%–10% for the night pack, and gourmet pack;
- 10%–20% for the cultural pack, cruise pack, and sea & tejo pack;
- 20%–25% for the caves pack.

Besides that, we establish a basic price between 10€ and 100€ for the gift voucher. This high discrepancy is justified by the existence of several different gift vouchers, which let our company to adapt to different consumer profiles.

<u>Promotion Policy</u>

In order to promote our services, we will focus on direct-action advertising and indirect-action advertising. The first approach intends to seek a quick response to the advertising message from the market; the second approach is essentially designed to stimulate demand over a longer period of time. Examples of direct-action advertising will be our e-mail marketing and social network presence; examples of indirect-action advertising are the advertising in buses and bicycles, participation in international trades, and establishment of partnerships.

Our main promotional channel will be the social network. We will use intensively four social networks to promote our services:

- Facebook – divulgation of our services and direct interaction with customers;
- Instagram – promotional images of our services;
- YouTube – promotional videos of our services;
- Google+ – creation of interest circles and sparks (contents' suggestions).

Additionally, we intend to promote our products using touristy blogs, Tripadvisor.com and booking.com, which one of the main websites used by tourists to plan their holidays and vacancies.

The establishment of partnerships will also be important to have a more versatile and complete packs and will also let potential customers to discover our services even if they do not know your website.

<u>Place Policy</u>

The main channel through which our services reach the final user will be our website and social networks. Therefore, we will use mainly direct channels. Using it, the customers can order the tourist card and check all relevant information related to packs, touristy routes, and discounts.

There are also intermediaries in our distribution channels, namely by using agents, tourist operators, and hotel facilities. Some of these intermediaries will be Travelplan, Nortravel, Soltur, AccorHotels, and InterContinental Porto.

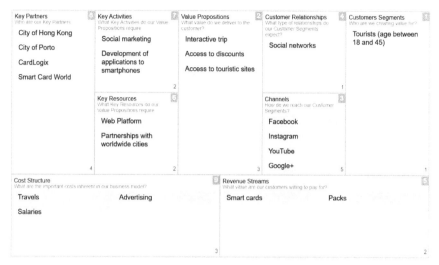

Figure 2.6 Business model canvas of TourMCard.

2.2.10 Business Model Canvas

The business model canvas is presented in Figure 2.6. *TourMcard* offers its customers an interactive travel experience with access to discounts and touristy sites. For this, it is essential to establish relationships with customers through social networks and the provision of a technological application with Web and mobile component. Our key clients are tourists between 18 and 45 years who visit the most tourist cities worldwide. It is essential to establish partnerships with the municipalities of these cities and also with suppliers of smart cards. The main revenues come from the sale of smart cards with personalized information for each city, the various tour packages associated and adapted to each city. On the other hand, and as the main costs, we have the necessary trips for the establishment of these partnerships, the costs of salaries and the advertising expenses to promote our company.

2.3 Scenario II – AuditExpert

2.3.1 Market Research

In order to estimate the market size and needs in the sector of IT, auditing will adopt both qualitative and quantitative approaches.

Starting by a qualitative approach, we identified the following relevant documents in the literature:

- Worldwide Security and Vulnerability Management 2013–2017 Forecast and 2012 Vendor Shares made by IDC Consulting;
- Global Cloud Security Software Market (2010–2014) performed by TechNavio Insights.

Based on these two documents, it was possible to extract the following conclusions:

- Market size estimated for 2014 is US $963.4 million with a growth rate of 41.4% CAGR (TechNavio Insights, 2014);
- Key geographies are the Americas, EMEA, and APAC (TechNavio Insights, 2014);
- Key customer segments include large enterprises, government agencies, cloud service providers (CSPs), and medium-sized enterprises (TechNavio Insights, 2014);
- Major trends in the market are in terms of increasing partnerships between CSPs and security solution providers, increasing emergence of cloud services for SMBs market and the emergence of strong cloud security standard and guidelines (TechNavio Insights, 2014);
- Key customer needs identified in the market belong to the following four areas: strong overall security, suite of security solutions, encryption key management features, and availability of fine granular control (TechNavio Insights, 2014);
- Top five vendors in the market are IBM, HP, EMC, Symantec, and McAfee (IDC, 2012).

In terms of a quantitative approach, we conducted an exploratory research. The survey is composed of 13 questions that intend to capture the need of an IT audit platform in a business environment. The survey was created in Google Drive and made available between 1 November 2016 and 31 December 2016. The survey is composed of 13 questions:

1. What is the dimension of your company?
2. What is your role in the organization?
3. Did you face any security incidents?
4. In your opinion, has the number of threats in the last 3 years increased?
5. What was the impact of data breaches in your organization?
6. Does your organization adopt any security policy?
7. What is your level of awareness about security in your organization?

8. Does your company adopt open standards?
9. Does your company adopt open-source software?
10. Do you use an IT audit platform?
11. Are you interested to have an IT audit security platform?
12. How much are you available to pay for it?
13. What are the main buying criteria?

It is also relevant to highlight that in a real business scenario it is important to present the results of the survey, the distribution of the respondents by each question, and its corresponding percentage. These detailed elements were omitted in the context of this book.

The results of the survey demonstrated that an increased number of SMBs are awareness about the need of IT security services, and most part of them is migrating for unified communication solutions. The number of IT threats has been increasing, particularly in terms of data loss, external threats, and denial of service attacks. Furthermore, the existence of open standards is considered mandatory in the IT security field and open-source software is gaining significant importance in the market. Finally, the main three buying criteria of our customers are: efficiency, integration with other IT tools, and price. The majority of our respondents stated that an IT security solution will be paid between 1000 and 2000 €.

2.3.2 Environmental/Industry Analysis

The PEST analysis is given in Table 2.14. In fact, the majority of elements addressed in PEST analysis is common between the two business examples.

The **political dimension** of the EU legislation is an aspect that may condition our organization. Our technical solution must be adapted to the appearance of new rules. Also, the emergence of cyber-terrorism may affect government and big companies. In this scenario, concerns about the security of information may arise. In **economic dimension,** the difficulties in credit access could potentially be one the biggest threats that may condition potential investments in the acquisition of IT solutions. In **social dimension,** the existence of significant difficulties to find qualified human power in IT security could have impact on salaries. Additionally, the rises in employee mobility turn difficult to guarantee high rates of job safety. Finally, in **technological dimension,** it is possible to realize the existence of three paradigms that are very important in this business field: open standards, open-source software, and cloud-based services.

Table 2.14   PEST analysis of AuditExpert

Political	Economic
- Political instability in several European countries - Weak intellectual property protection in emerging markets - Increase importance of EU legislation - Emergence of cyber-terrorism	- Exchange rate fluctuation - Economic stagnation - Rising energy costs - Difficulties in credit access
Social	Technological
- High employment rate - Difficulties to find qualified human power in IT security - Rise in employee mobility - Low illiteracy rate in developed countries	- Easy access to media - Easier access to the Internet with broadband connections - Increasing importance of open-source technologies - Adoption of open standards in the IT field - Lack of proper cloud security standards - Mobile and cloud-based services are gaining more importance - Increasing availability of cloud security solutions for SMBs

After the PEST analysis, we will look also for the Porter's Five Forces that is given in Table 2.15.

The **threat of new entrants** is **moderate** due to several antagonistic characteristics. From one perspective, we can realize that it is not easy to introduce new entrants in the market essentially due to the big learning curve and high protection mechanisms, some of them based on patents. However, because it is easy to implement differentiation in the products and services and also because the industry has a high level of attractiveness, the appearance of new players emerges as a real threat. The **bargaining power of buyers** is **moderate** because it is moderately easy to lose buyers if a new competitor appears in the market; the **bargaining power of suppliers** is **low** because it is very easy to find new suppliers in the market, particularly in the open-source domain; the **threat of substitute products** is **high** due to the emergence of cloud and mobile services that have potential in the paradigm of IT security audit industry; finally, **the rivalry among existing competitors** is **moderate** because the number of competitors in the market is still low compared to other industries but they typically operate in several market segments.

Table 2.15 Porter's Five Forces analysis of AuditExpert

Threat of new entrants	- Big learning curve
	- High protection mechanisms
	- Easy to offer product differentiation
	- Difficulties to enter as a local player
	- Industry growth rate is high
	- Need of very specialized knowledge in IT
Bargaining power of buyers	- Number of buyers are low
	- Switching costs are moderate
	- Price sensitivity is low
Bargaining power of suppliers	- Many alternatives for suppliers
	- Most of the companies in the market develop also their own technologies
	- High availability of substitute technologies
	- Costs of changing are low
Threat of substitute products	- Emergence of cloud services
	- Emergence of mobile services
Rivalry among existing competitors	- Fixed costs are high
	- Increase diversity of competitors
	- Low segmentation
	- Customer loyalty is still high
	- Rivalry among large diversified players for acquisition of niche players

2.3.3 SWOT Analysis

The strengths are essentially concentrated in two dimensions: technology and people organization. On the one hand, the fixed costs are low due to the utilization of open-source technologies. Additionally, the flat organization and the characteristics of start-up companies make us more reactive and adaptive to market changes. On the other hand, the weaknesses are also related to the reduced size of our company. The low experience of our team in this specific market can bring additional issues, especially in the short-term period. Besides that, we expect to have difficulties to penetrate in big customers and a severe competition from incumbent players. The opportunities offered by the market are substantial. There is a high potential of market growth; it is relatively easy to use the Internet and our partners as important distribution channels without having international physical offices, and the appearance of new technologies can reduce significantly the development and maintenance costs. Finally, some threats should be properly mitigated. For example, these new technologies that could appear can also be explored by

Table 2.16   SWOT analysis of AuditExpert

Strengths	- Low fixed and distribution costs - Flat organizational structure - Experience of the team in the field of computer networks - Ability to personalize the service to different customer segments - High availability of services (24/7) - Use of open-source technology - The solution provides granular fine detailed control, providing a policy-based approach to key management and data access
Weaknesses	- Low experience of the team in the IT security market - Lack of brand image - Lack of experience of the management team - Difficulties to penetrate and get market share from big players in the market - Limited geographic presence - Low established customer base that typically the traditional security solution vendors enjoy
Opportunities	- High potential of the market - High degree of customers' loyalty - Access to international markets - Appearance of new distribution channels - Appearance of new technologies - Synergy with associated partners
Threats	- Appearance of new technologies - High dependence of partners and negative performance of partners - Threat from bigger established traditional players in computer networks that have better geographic reach and that may therefore expand more quickly

our direct and indirect competitors, and we expect to have a high dependence of our partners. The SWOT analysis is provided in Table 2.16.

2.3.4 Customers Identification

There are generically four segments in the field of IT audit security: (i) large enterprises, (ii) government agencies, (iii) cloud service providers (CSPs), and (iv) small and medium-sized businesses (SMBs).

Large enterprises employ 250 or more people (OECD, 2016). Companies that belong to the group of large enterprises have grown to the point where it needs dedicated, full-time IT staff with specific expertise to manage specific applications or parts of the IT infrastructure. Typically, they have specific IT roles such as systems administrators, exchange administrator, database administrator, and helpdesk staff. They also traditionally have subcontracting services with consulting firms.

Government agencies are characterized as administrative units of government that are tasked with specific responsibilities. These agencies can be established by national, regional, or local governments. Any given government is likely to have hundreds of agencies with a variety of objectives and roles. These entities are similar in a lot of areas to large enterprises, but typically they have a high number of legacy software. This situation turns more difficult to manage an integrated IT solution.

Cloud service providers (CSPs) are companies that offer network services, infrastructure, or business applications in the cloud. Therefore, the cloud services are hosted in remote data centers that can be accessed by companies or individuals using network access. The largest benefit of using a cloud service provider comes in efficiency and economies of scale. Rather than a company building its own infrastructure to support internal services and applications, the services can be acquired from the CSP, which provides the service to many customers from a shared infrastructure.

Currently, CSPs store services for critical data storage and there is an increase in the number of cloud-specific attacks. Despite the strong growth drivers and interesting trends, CSPs market has also some growth inhibitors. Some CSPs still believe that cloud security is only the end-user's responsibility, nor is security a strong buying criterion for end-users. Hence, some providers are not particularly interested about the offering of cloud service solutions. Another major issue is the lack of awareness among end-users about the risks associated with cloud computing. Other challenges are inconsistent network connections and lack of proper cloud security standards.

Small and medium-sized businesses (SMBs) are characterized to have fewer than 250 employees (OECD, 2016). There are slight differences in this definition with the introduction of micro and SMEs, and limits to the annual turnover (i.e., not exceeding 50 million euro). SMBs only have few dedicated IT staff that, in some cases, can only be part-time individuals. This staff does typically everything, such as manage backups, databases, the network, support contracts, etc. SMBs are slowly adopting cloud security solutions, and the most prominent adopters are companies from financial services and healthcare, retail, and technology sectors.

2.3.5 Competitors Identification

Table 2.17 identifies the main competitors in the market. We identified two groups of competitors: (i) unified communication solutions and (ii) IT security audit companies.

Table 2.17   Competition overview of AuditExpert

Competitor	Activity Start Date	Market Share (%)	Strong Points	Weak Points
Critical Links	2006	Not available	- Unified Communications Solution or "Office-in-a-box" that provides all the phone and Internet communication services a small and medium-sized business needs - Comprises a Stateful Inspection Firewall, VPN for remote user access (PPTP), or site-to-site (IPSec) - It is based on open standards	- Low flexibility of the platform that does not have a modular architecture as IPBrick - Security protection is relatively basic - Scalability to be used in large companies is low
IPBrick	2001	Not available	- It offers a low TCO (total cost of ownership) - Disaster recovery in just 15 minutes - It is based on open-source technologies and open standards - Possibility to integrate award-winning security tools	- Security protection is relatively basic - Difficulties to integrate proprietary software

Table 2.17   Continued

Competitor	Activity Start Date	Market Share (%)	Strong Points	Weak Points
Trend Micro	1988	13–17%	- The company is the leader in both the cloud security and virtualization security markets - The company has partnerships with leading IT organizations such as HP, Cisco, Dell, Microsoft Corp., Oracle, and Wipro, which can be leveraged to expand its operations - It has released cloud security products for SMBs, which are expected to be strong adopters of cloud security	- Company's overall revenue growth is below the annual average growth of global security software market - Around 40% of the revenue comes from the Japanese market - The solutions provided by some of the pure-play companies are stronger
McAfee	1987	8–10%	- The company's cloud access control solution allows control over the entire life cycle of cloud access security - The company has a very long list of partners from various geographic locations, which it can leverage to expand	- Difficulties to have enough flexibility to face the SMB market - Some of its competitors, such as Trend Micro, provide better key management options/features - Threat from pure-play vendors
Symplified	2006	6–8%	- The company is the market share leader in the cloud identity management space - Its solution supports different smartphone operating systems such as iOS, Android, and BlackBerry It has recently introduced Symplified Mobile Edition that secures cloud and Web applications on any mobile device	- While the company is strong in the cloud identity management space, it is not as strong as in other spaces such as application security, encryption, and access control - It has limited geographic presence and draws the majority of its revenue from the Americas

(*Continued*)

Table 2.17   Continued

Competitor	Activity Start Date	Market Share (%)	Strong Points	Weak Points
			- It has partners such as Amazon Web Services, Google, Salesforce.com, Cloud.com, and Wipro, that can be leveraged to expand further	
Black Duck	2003	Not available	- Offices in several international markets - Competencies in identifying security threats in open-source code - Realization, on demand, of open-source security audits - Security audits focus mainly on software domain	- Lack of knowledge in property solutions - Does not offer an integrated unified communications solution

Below we provide detailed information about each competitor.

Critical Links

Critical Links is focused on providing innovative IT solutions for common problems faced by schools and companies around the world. EdgeBOX is a recognized unified communication solution provided by Critical Links. The edgeBOX provides a full business phone system (IP-PBX) in a single appliance that can be managed remotely through an easy-to-use interface.

EdgeBOX is based on robust software providing a highly available and reliable system. Since it is a fully integrated "all-in-one" device, there are fewer devices and interfaces than can fail. If edgeBOX fails, time to locate and rectify the fault is reduced and a replacement unit can be swapped in and restored from automatically produced backup files. edgeBOX is based on open standards and has a flexible architecture that enables new services to be added while the device is still operational. Dedicated, enterprise class hardware provides the platform for edgeBOX software, and it can be easily upgraded to provide additional capacity for new users or services.

EdgeBOX provides the following features:

- Access router – manages the traffic between your company LAN, the Internet, and a DMZ (demilitarized zone). Provides all basic network access services such as DHCP, NAT, DNS, and VLANs;
- File storage – provides optional redundant disks (RAID) for your critical business data;
- IP-PBX – provides all the telephony features small business needs, including sophisticated call conferences, parking and forwarding, IVR, LCR, ACD, fallback to PSTN, etc. The full featured IP-PBX supports both VoIP and PSTN calls;
- NAC – Network Access Control allows you to control how employees and visitors access your network. Access can be granted/denied per user. NAC supports authentication through a captive portal or 802.1X;
- Security – secures your company network against attacks or misuse. Comprises a Stateful Inspection Firewall, VPN for remote user access (PPTP), or site-to-site (IPSec). Authentication at the firewall allows tight control over user activity in the network (who, when, what);
- Quality of Service (QoS) – guarantees that your priority services have the bandwidth available when needed, e.g., for ensuring business quality calls using VoIP and lowering the priority of other traffic such as Web downloads;
- Management – provides a unique and easy-to-use Unified Management Interface that allows the configuration of almost all aspects of the system by non-IT experts. Processes such as adding a new user typically involve configuring a multitude of devices and services take just a few minutes through the edgeBOX Management interface.

IPBrick

IPBrick, S.A. is a company that has the mission to constantly innovate in communications solutions for companies. IPBrick solution is composed of four modules: one core module (IPBRICK OS) and three appliances (IPBRICK.GT, IPBRICK.I, and IPBRICK.SOHO).

IPBRICK OS is a communications platform for enterprise server systems based on Linux Debian, and using the most-known open-source software to manage unified communications, document & process management, email & groupware, and enterprise social network. IPBRICK OS also provides a high-level integration between the pre-configured open-source software packages. The integration is not limited to only IPBRICK OS services, but is also extended to third-party software applications, which are IPBRICK OS ready.

IPBRICK.GT is a pioneering implementation of UCoIP (Unified Communications over IP). UCoIP includes voice, video, fax, email, SMS, IM, and Web as communications core of every business. UCoIP adds mobility and enterprise communications integration to the telecommunications world. It fits the customer needs with full flexibility and the lowest TCO (total cost of ownership).

IPBRICK.I is the intranet component for IPBrick solutions. Intranet is the application of Internet technologies into companies' and organizations' internal networks, improving the business processes with low costs. This makes it possible for all employees to have better access to strategical information inside the company, also reducing operational costs. IPBRICK.I is based on open-source technology and integrates every kind of software, enabling a short-time installation.

IPBRICK.SOHO is a solution providing in one equipment two essential tools in every office – PBX and fax machine. Furthermore, it also provides email, SMS, Web server, and instant messaging, all integrated in a single appliance. It also provides a full set of proxy services to prevent workstations from connecting directly to the Internet.

Trend Micro

Trend Micro Inc. is a major security solutions provider. The company was incorporated in 1988, is listed on the Tokyo Stock Exchange, and is headquartered in Tokyo, Japan. It has offices in North America, Latin America, the EMEA region, and the APAC region. It has operations in 23 countries around the world. The company serves several sectors of activity in the home office, small businesses, medium-sized businesses, enterprises, and service providers. It has a strong partnership network, and in its partnership ecosystem, Trend Micro includes consulting and service partners, managed services and hosting partners, platform partners, technology partners, and channel partners/resellers.

The company reported revenue of US $1.085 billion in FY2010. The company draws more than 60% of its revenue from business customers and the rest from individual consumers. The company drew 41% of its revenue from Japan in FY2010. North America accounted for 26%; Europe, 21%; the APAC region (other than Japan), 9%; and Latin America, 3%. However, revenue from the Cloud Security segment was only around US $30 million. The majority of the cloud security revenue is expected to come from the Americas and the Europe region.

McAfee

McAfee Inc., founded in 1987, is a leading provider of network security solutions. The company is listed on the New York Stock Exchange and is based in California, US. The company has categorized its geographic operations into five segments: North America; the EMEA region; the APAC region excluding Japan; Japan; and Latin America. The company's security products are for sectors such as Data Protection, Email and Web Security, Endpoint Protection, Mobile Security, Network Security, Risk and Compliance, Security Software as a Service (Security SaaS), and Security Management.

At the end of 2010, the company reported total revenue of US $2.1 billion. Product revenue was US $225.4 million, whereas services revenue was US $1.8 billion. North America contributed the maximum, 58%, to revenue. The EMEA region followed with 26%. Japan came next at 7%, and the APAC region and Latin America contributed 6% and 4%, respectively. However, in 2010, the revenue from the Cloud Security segment was low, at only around US $25 million.

Symplified

Symplified Inc., founded in 2006 and based in the United States, is a leading cloud identity management provider. The company serves industries such as cloud providers, energy and utilities, financial services, healthcare, high tech, and life sciences. The company categorizes its solutions into access control, authentication, audit and provisioning, and administration. It provides solutions such as Symplified Access Manager, Symplified Identity Manager, Symplified Sign-On, and Symplified SinglePoint Platform-as-a-Service (PaaS). Symplified has a strong list of technology and channel partners. It has technology partners such as Amazon Web Services, Digipass, Google, Salesforce.com, Taleo, and Vasco. Further, the company has channel partners such as Cloud.com, Cloud Distribution, Computer Sciences Corporation (CSC), Kumoya Inc., and Wipro Technologies.

Symplified is a private company and does not report its revenue. However, the revenue from cloud security is expected to be in the range of a few million US dollars only. The major contribution is expected to come from the Americas. It is believed that the company draws the majority of its revenue from its product segment.

Black Duck

Organizations worldwide use Black Duck's industry-leading products to automate the process of securing and managing open-source software, eliminating the pain related to security vulnerabilities, compliance, and operational risk. Black Duck is headquartered in Burlington, MA, and has offices in San Jose, CA, London, Frankfurt, Hong Kong, Tokyo, Seoul, Vancouver, and Beijing.

The Black Duck software provides a deep scanning. Hub provides full source and binary scanning that identifies all open source (even modified code) used in apps & containers and maps vulnerabilities using multiple databases. Black Duck Hub integrates directly into company environment and processes, enabling those organizations to find and fix vulnerabilities at every stage of the development life cycle. This approach reduces development time and cost, while providing protection against open-source risks.

2.3.6 Partners Identification

There are two kinds of strategic partners for our business. In the first group, there appear companies that provide unified communication solutions where their core business is not in the field of IT security. In the second group, we find companies that already provide IT audits, but they still have significant margin to increase their audit robustness in IT security field. Last, but not least, it is important to highlight that some of these companies were also identified as potential competitors. This situation is common in the strategic management field where our positioning is flexible to adapt our positioning in function of our partners and competitors. Table 2.18 provides an overview of our potential business partners.

2.3.7 Strategic Positioning

AuditExpert intends to have a strategic positioning where we target mainly the establishment of partnerships instead of end customers. Therefore, we plan to have a conservative strategy to end-user markets and a more aggressive positioning in identifying and establishing partnerships with our potential partners.

There are also several factors that are mandatory for our company. First, it is very important to have a distinctive IT solution that could be highly robust in providing IT security, but simultaneously flexible enough to be integrated with open-source and property solutions. Second, it is fundamental to have

Table 2.18 Partners overview of AuditExpert

Partner	Activity Start Date	Size	Market Share (%)	Benefits for Our Company	Benefits for Partners	Potential Issues
IPBrick	2001	51–200 employees	Not available	- Significant number of customers in Portuguese and European markets - Integration of our IT security solution in a unified communication solution	- Reliable security systems in a full-of-box IT solution - Realization of IT security audits	- Integration between open-source and property software - IT audits is not the core business of IPBrick
KPMG	1987	More than 10,000 employees	Not available	- Benefits from experience of KPMG in the global audit market	- Extension of KPMG business in a high growth market	- Lack of experience of KPMG in IT industry - High bureaucracy to establish a partnership with KPMG - Significant differences in the organizational structure between two companies
Protiviti	2002	1000–5000 employees	Not avilable	- Proved experience in IT audit from our partner - Access to a high variety of customers	- Specializing in IT security field	- Protiviti can offer alone IT security specialized solutions in the near future

(Continued)

Table 2.18   Continued

Partner	Activity Start Date	Size	Market Share (%)	Benefits for Our Company	Benefits for Partners	Potential Issues
OpenVAS	Not available	Not available	Not available	- OpenVAS provides a framework of several services and tools for vulnerability management and scanning - Can help us to improve our knowledge and competencies in IT security field	- Migration from products market to services field - Realization of IT security audits	- The core competencies of OpenVAS is only in the products field - Lack of experience in consulting services
Zone Minder	Not available	Not available	Not available	- Deep experience in IT monitoring field - Experience in the integration of open-source software with property solutions - Access to a high variety of customers	- Extension of their business activities to IT security and audits	- The focus of ZoneMinder is only in surveillance software systems

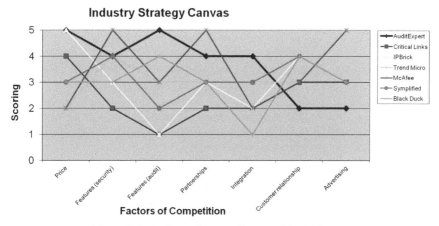

Figure 2.7 Industry Strategy Canvas of AuditExpert.

a total customer relationship solution that could help us to establish a high degree of collaboration with our partners.

Figure 2.7 provides a comparative vision of our strategic positioning in relation to our competitors.

Our company will adopt a niche marketing strategy. We plan to focus on providing an IT security solution with access to audit security report. We will intend to provide an offer that can be characterized by the following attributes:

- Low price – due to the reduced level of functionalities provided by the application, which focus only in providing IT security and IT audits;
- High level of available features in terms of security and IT audit – this is possible due to the absence of features in the field of computer networks and unified communications, which is outside of our scope;
- Quality of our partnerships – instead of having a high number of partners, we intend to strategically choose our partners in order to keep a strong and cooperative relationship with them;
- High level of integration capabilities – in order to facilitate the integration of our solution in a full-out-of-box IT communications solution provided by our partners.

2.3.8 Advertising & Sales

Table 2.19 presents the advertising channels that will be used to promote our product and services.

Table 2.19 Advertising channels

Advertising Channel	Benefits	Costs
Social Networks	- Increasing marketing force - Increased knowledge of the company - Better interaction with current and potential customers	Not available
E-mail marketing	- Intensification of the company advertising - More direct marketing and connected to the characteristics of the customer - Personalization of the message to specific customers - Knowledge of updates, promotions, and other benefits for customers	Not available
Partnerships	- Strengthening confidence in our brand - Increase connection between our company and partners	Not available
Participation in international fairs	- Establishment of partnerships - Get contact with corporate customers	1600–3000 €
Participation in job fairs	- Divulgation of business activities - Establishment of partnerships with universities - Recruitment of IT qualified employees	400–800 €
Word-of-mouth campaigns	- Divulgation of product and services - Personalization of message to direct customers	Free
Endorsement campaigns	- Divulgation of product and services	Not available
Technical magazines	- Intensification of the company advertising - Increase brand image	500–1000 €

There are three situations that are relevant to highlight:

- Participation in international trades includes two presences in international fairs by year. It includes the registration price and logistic costs;
- Participation in job fairs to increase our interaction level with universities' stakeholders. This type of collaboration will allow us to attract qualified IT employees to work for our company;

Table 2.20 Distribution channels

Distribution Channel	Product/ Service	Percentage of Sells	Advantages	Disadvantages
Website	- Sells of security IT application - Realization of IT security audits	25% for products ; 100% for services	- Product will be installed in client within 24 hours - Easy to use the services - Availability 24/7 hours/days	- No physical evidence of the product
Partners	- Sells of security IT application	75% for products	- Support from our partners	- Extra cost - Delay until 72 hours of product installation

- Endorsement campaigns consist in finding a key influencer person in the IT security market. This person will make a presentation to potential customers, highlighting the added value and benefits offered by our product.

Only two distribution channels will be used. For product sells, our customer can place an order on our website, or using our partners. For service sells, only the website can be used. Table 2.20 details our distribution channels.

2.3.9 Marketing Mix

Product/Service Policy

Our product policy is characterized by the existence of one product and one service. Our product is the IT unified security solution that offers the following features:

- Communication and messaging – companies can access and manage email, voice, and fax from a single interface;
- Platform and infrastructure – centralization and simplification of the technology infrastructure to deliver high performance, flexibility, and agile solutions;
- Video and conferencing – deliver high-performance video solutions, with great quality and error resiliency;

Table 2.21 Price Policy

Number of Users (max)	Connectivity Options	Storage	Price
10	- WiFi access point - Up to 5 analogue lines - Up to 5 ISDN2 lines - Up to 2 E1/T1 lines - GSM/3G support	250 GB	1000 €
50	- Wi Fi access point - Up to 10 analogue lines - Up to 10 ISDN2 lines - Up to 5 E1/T1 lines - GSM/3G support	500 GB	2500 €
250	- Wi Fi access point - Up to 20 analogue lines - Up to 20 ISDN2 lines - Up to 10 E1/T1 lines - GSM/3G support	1 TB	5000 €
No limit	- Wi Fi access point - Up to 30 analogue lines - Up to 30 ISDN2 lines - Up to 15 E1/T1 lines - GSM/3G support	2 TB	10,000 €

- Communication security – customized security solutions in terms of firewall, VPN, IDS, anti-virus, anti-malware, and anti-spam.

The IT unified security solution is expected to have a significant impact on the SMB market, by improving efficiency, reducing costs, increasing revenues, and offering a better customer service.

Our customers' client has also access to an audit security report that offers a premium service to our clients. The IT audit is performed in a non-intrusive way by the adoption of agents that will audit the IT security network, Web applications, and mobile solutions. The audit performs a risk assessment analysis and proposes a personalized mitigation plan.

Price Policy

The price of our IT unified security platform has a significant range that depends on the maximum number of users, connectivity options, and storage space. This situation is depicted in Table 2.21.

The price of the IT security audit is calculated according to the scope of the audit and the size and complexity of the company infrastructure. There is also a basic minimal price for the IT security audit is 1000 €. A budget is given to the customer before any audit is performed.

<u>Promotion Policy</u>

Our promotion policy is based on the existence of strong partnerships. These partnerships will allow that a significant amount of our sales comes from our partners. The idea behind this strategy is to let our partners use our product in a more wide-scope IT network management platform that could be used by big enterprises.

However, and in order to reach the SMB target market, we will also use other promotion strategies in order to increase the awareness in our products and services. Therefore, we plan to have regular participation in international trades, ads in technical magazines, and the use of endorsement campaigns.

Finally, we will adopt direct channels to interact and communicate with our customers by using social networks and e-mail marketing campaigns.

<u>Place Policy</u>

Around 75% of your products sells will expect to come from our partners that will be responsible to sell directly or incorporate our IT unified security solutions in their network communication solutions. Besides that, we expect to receive orders from SMB market directly on our website that incorporates an eCommerce application developed using Magento.

Finally, requests about realization of IT security audits are performed on our website. Like in the previous scenario, our eCommerce solution will register these requests and billing all the process.

2.3.10 Business Model Canvas

The business model canvas of *AuditExpert* is provided in Figure 2.8. We emphasize first the foremost key partners of the company that we consider fundamental to the success of the business. We also have the main activities of the company, which consists of having a data center that allows managing all the applications installed in the clients, the audit process, and also the activities of marketing and selling. To this happen, it is necessary to invest in quality certifications in the security field and a robust and efficient technological infrastructure. We also emphasize that we have very diversified panoply of customer segments, in which we intend to establish relationships through partners and social networks. Partnering is a crucial part of our business, as it

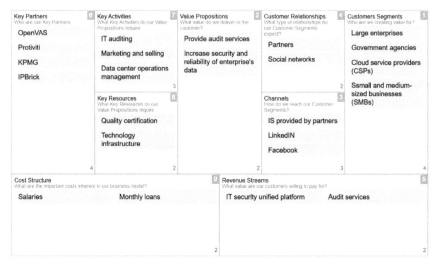

Key Partners Who are our Key Partners	Key Activities What Key Activities do our Value Propositions require	Value Propositions What value do we deliver to the customer?	Customer Relationships What type of relationships do our Customer Segments expect?	Customers Segments Who are we creating value for?
OpenVAS Protiviti KPMG IPBrick	IT auditing Marketing and selling Data center operations management	Provide audit services Increase security and reliability of enterprise's data	Partners Social networks	Large enterprises Government agencies Cloud service providers (CSPs) Ssmall and medium-sized businesses (SMBs)

Key Resources
What Key Resources do our Value Propositions require

Quality certification

Technology infrastructure

Channels
How do we reach our Customer Segments?

IS provided by partners

LinkedIN

Facebook

Cost Structure What are the important costs inherent in our business model?		Revenue Streams What value are our customers willing to pay for?	
Salaries	Monthly loans	IT security unified platform	Audit services

Figure 2.8   Business model canvas of AuditExpert.

is through it that we aim to reach a very diverse and wide range of customers. The structure of costs stands out the salaries of highly qualified professionals and monthly loans that is the result of an initial investment through sources of external financing. On the revenue side, IT security unified platform and audit services in the domestic and international markets stand out.

3

Operational Plan

Overview

The operational plan is concerned with the day-to-day function of running a business. It focuses the analysis on the administrative and production processes, which will help the company to improve the quantity while reducing costs. This chapter begins by presenting the company's production process, which describes the involvement with suppliers, the necessary material, payment process, and quality control policy. Additionally, a plan of action is presented in which a set of milestones are defined with the identification of the person/role responsible for its execution. Finally, we present the operational plan of *TourMCard* and *AuditExpert*.

3.1 Theoretical Foundations

3.1.1 Operations

In terms of operation, the entrepreneur must think in the following components of the business:

- Production process – what process is involved in producing your products or services? Notice that this process may vary depending on your product or service;
- Suppliers – who are your main suppliers? what do they supply to your business? how will you maintain a regular good relationship with them?;
- Plant and equipment – list the current plant and equipment purchases, which can include vehicles, computer, industrial equipment, communication devices, etc. For each equipment, it is important to register the purchase date, purchase price, and yearly running cost;

- Inventory – list of the current inventory items. If there is a substantial inventory list, it is preferable to include it in the annex. For each inventory item, it is important to register the unit price, quantity in stock, and total cost;
- Software – what kind of software is needed (e.g., for website, point-of-sale software, or accounting package)? what will be the main purpose of each? will they be off-the-shelf, or purpose built? what is the estimated cost of each technology solution?;
- Communication channels – how can your customers get in contact with you? These channels can include mobile phone, email, fax, and/or Internet blog/social networking website;
- Payment types accepted – what payment types will you accept (e.g., cash, credit cards, gift cards, paypal, etc.)?;
- Credit policy – what is your credit policy for customers/suppliers? how long is the credit period? what credit does your business receive?;
- Quality control – what is your quality assurance strategy? what steps do you take to meet the product safety standards?

The **production process** is characterized to have several set of iterative activities that can be modeled using a workflow. Along the production process, the resources are combined and transformed in a controlled manner and in accordance with the policies communicated by the management team. Manufacturing operations can be broadly divided into two categories:

- Made to stock production process – produce items that are stored as stock, before the customer receives it;
- Made to order the production process – the production process starts only after receiving an order from a customer. It is based on the principle that the manufacturer cannot anticipate what each customer needs.

There is also another type of production process. This third type of production process is called the "Assemble-to-Order" production process. In this model, the company produces standard modules and assembles them accounting for the specification of customer orders.

These production models presented above are traditional and common to most sectors of activity, but in the IT field, we need to describe it using specific language and tailored to software development models. There are several software development processes; however, the traditional waterfall model contemplates several basic sequential activities such as requirements gathering, requirements analysis, design, implementation, testing, and deployment. In recent years, other models of software design have emerged, with particular emphasis on agile methodologies. They emerged as a response to the bureaucracy of traditional engineering methodologies. The main differences brought about by agile methodologies is that they are less document-oriented, more adaptive than predictable, and people-oriented rather than process-oriented.

The **suppliers** are the persons or companies that provide something to another company or community. The relationships that companies have with suppliers enable them to gain competitive advantage over their competitors by offering more value to their customers. In this way, suppliers can be seen as a group of interests that provide them with the inputs needed to produce goods or services. A supplier has a high importance within every industry sector, and IT field is not an exception. Suppliers are not limited to providing goods and services, but they have emerged as an asset in business relations.

The supply management channel has emerged as a very relevant area in the activity of an organization, because this management can guarantee the best products/services in the market with the best conditions, timely delivery, among other advantages. Poor supplier management with recurrence of suppliers who have failed previously or who have not complied with the contract can compromise the entire operation of the organization. Therefore, it is important to gather different objective metrics for the evaluation of the suppliers and establish key performance indicators (KPIs).

Suppliers in the IT field continue to exist like in other activity sectors, especially the need of specialized equipment and/or software. The use of open-source software is also an emerging paradigm that changes the way products are developed. Open-source software can increase the quality of developed software and provide lower costs, but raises other important issues to consider, such as sustainability, accountability, and monetization.

The **plant and equipment** are responsible to present the list of equipment that may be needed for the company's operation. All the equipment in the list must have a purchase and running cost. It shall be described what does each equipment, how the parts work together, and how much they can produce. In this section, the work equipment, vehicles, computers, and office equipment shall be included.

The **inventory** is basically composed of a list of goods and materials available in stock that are stored inside the company or stored externally but owned by the company. Available materials listed in an inventory can be used in the manufacture of more complex goods or else they can be marketed themselves, depending on the company's business. There are generally four types

Table 3.1   Stock levels management strategies

	Advantages	Disadvantages
No stock or low level of stock	- Efficient and flexible by guaranteeing that the company only owns what it is needed; - Lower storage costs; - Easy to keep up to date and develop new products without wasting stock.	- Meeting stock needs can become complicated and expensive; - The company can run out of stock as there is a hitch in the system; - High dependence on the efficiency of your suppliers.
High level of stock	- Easy to manage; - Low management costs; - Difficult to reach a situation of run out; - Buying in bulk may be cheaper.	- Higher storage and insurance costs; - Certain goods might perish; - Stock may become obsolete before it is used; - High needs of frontend capital.

of stock: (i) raw materials and components, (ii) work in progress – stocks of unfinished goods, (iii) finished goods ready for sale, and (iv) consumables.

Deciding how much stock shall be kept depends on the size and nature of your business and the type of stock involved. Table 3.1 summarizes the main advantages and disadvantages associated with two strategies of stocking levels. In service companies, it is common that there may be no inventory of products.

The use of no stock or low level of stock is suitable in a fast-moving environment where products develop rapidly. The stock is expensive to buy and store or the items are perishable. This strategy is particularly adopted in the IT sector. On the other hand, the adoption of high stock levels is suitable for businesses where sales are difficult to predict. Additionally, it is useful when there is a high level of price changes, and it is important to store plenty when the stock is cheap. This strategy is particularly adopted in the oil market.

The **software** section details information about the IT technology used by companies to operate. The software needed varies from company to company, but generically we can identify the following domains of applications:

- Accounting software – allows the management of payments and receipts. In addition, it allows the management of taxes and legal obligations;
- E-mail – allows direct communication with customers and suppliers. It is often the first gateway to the company;

- Documentation – used to write documents and produce reports or presentations;
- Spreadsheets – used to make a basic and daily financial analysis of the operations;
- CRM – used to make an integrated management of the customer relationship.

Each piece of the software bought by the company has an associated license file that stipulates what can be done with it. Essentially a software license contains two pieces of information: (i) the number of computers that can be installed and how many users can run it and (ii) the type of user permitted to use the software. These two elements are typically associated with the property software. However, the open-source software also has associated licenses and its use is even more complex. Basically, each license guarantees different rights and obligations in terms of use and distribution. It is important to guarantee that there is always compatibility between licenses. The most known open-source licenses are: GNU General Public License (GPL), GNU Lesser General Public License (LGPL), BSD license, MIT license, and creative commons.

The **communication channels** are the methods that a company uses to get in touch with the target audience in order to advertize their services or to follow a sales process and even solicit feedback from the customer. In addition to intensifying the sales effort, the communication channels help the company to consolidate its image and credibility in the market. The more efficient these communication channels are, the better will be the image of the company for customers. There is a common coincidence between the communication channels and the distribution channels. However, some can be different, since there may exist excellent channels of dialogue that are not responsible for big sales. They are channels that work well for prospecting, negotiation, relationship, and customer support. The most common communication channels are telephone, email, chat, complaint channels, and FAQs. In the IT field, all these channels can co-exist but with different levels of importance.

The **payment types accepted** must clarify which are the types of payments that the company accepts. The traditional payment in cash is still preferable in a wide range of situations; however, the advent of online commerce platforms has changed this paradigm. Other forms of payments, such as paypal or credit cards, have emerged and gained significant market share. If the form of payment can bring advantages and discounts to the customer, then it needs to be properly considered by the merchant. The point is not only to offer several alternatives to the customer, but also to study if the costs of these options cannot disrupt the business, making it less profitable. The most common payment types are cash, debit cards, credit cards, paypal, MB NET, bank transfer, and cash on delivery (COD).

The **credit policy** establishes a set of standards or criteria that each company uses to finance or lend resources to its customers and suppliers. Each company must develop a coordinated credit policy to find the right balance between sales needs and, at the same time, sustain a high quality. The credit policy aims to balance profit objectives with customer needs. The goal is to achieve a risk-adjusted profit objective and satisfy customers while maintaining a solid credit portfolio. Credit policies help achieve this balance and include rules that define appropriate behavior, standards, or performance criteria that measure compliance with trade policies and objectives, and procedures that define specific activities to ensure that standards are met.

Finally, the **quality control** is related to continuous improvement and standardization of processes and procedures. Their results should be measured, because they have a positive impact on product quality or service delivery, operational effectiveness, and financial performance of the organization. Typically the quality control serves many purposes, including improving processes, reducing waste, lowering costs, and engaging staff.

In the IT field, the quality of the software can be essentially analyzed through the level of the process maturity. Thus, the maturity of an organization's software processes influences its ability to achieve cost, quality, and schedule goals. Improving the software process offers a return on investment that should try to be measured, although there are additional benefits that are intangible and cannot be easily quantified. Among the most adopted quality standards in the software field, we have the CMMI, ISO 12207 and ISO 15504, also known as SPICE (Software Process Improvement and Capability Determination).

3.1.2 Action Plan

Action planning is the process that guides the day-to-day activities of an organization or project. It is the process of planning what needs to be done, when it needs to be done, by whom it needs to be done, and what resources or inputs are needed to do it. In fact, it is the process of operationalizing the defined strategic objectives.

The action plan must define the framework of the action against the defined strategy, the objectives to be achieved by this action, the person responsible for the action, as well as other actors and possible constraints to the development of the action. Finally, it should schedule the action and budget, indicating the expected results with it, which will serve as the basis for the evaluation.

Most action plans consist, at least, of the following elements:

- Milestone – presents what are the business milestones that you need to complete starting from current date;
- Date of expected completion – indicates when do you expect to complete them;
- Person responsible – indicates who is responsible for delivering this milestone.

3.2 Scenario I – TourMCard

3.2.1 Operations

Production process

The company adopts Scrum as the agile methodology for software development. According to the guidelines of the "Agile Manifest", people play a key role in the development of software projects. Therefore, it is essential that exists a good communication between the stakeholders, motivation, and that each individual is concerned about the quality.

The Scrum methodology employs an iterative structure and incremental in the following way: at the beginning of each iteration, the team analyzes what should be done and then chooses what they believe may become an increment of the value to the product at the end of the iteration. The team does its best to carry out the development of that iteration and at the end presents the increment of built functionality so that the stakeholders can check and request changes at the appropriate time.

In Scrum, a project begins with a simple view of the product that will be developed. The list of functionalities is presented and prioritized in the product backlog. All work is divided into Sprints, which are iterations of 2 weeks. Every day, the Scrum Team meets in a 15-minute meeting called the Daily Scrum to synchronize the whole team's work and inform Scrum Master of any impediments at work. After the Sprint Review Meeting and before the next Sprint, the Scrum Master meets with the Scrum Team in a last meeting: a retrospective meeting. Figure 3.1 illustrates the Scrum adopted methodology.

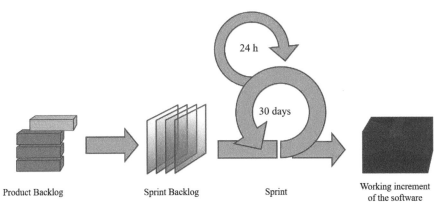

Figure 3.1 Scrum methodology.

Due to the reduced size of our development team, the role of Product Owner and Scrum Master for 2017 will be performed by the same person. Then, in 2018, we will intend to have these two separated roles.

Suppliers

The company has two suppliers (CardLogix and Smart Card World) that are responsible for the production of seamless cards. In order to decrease the products cards, we will provide always a batch order of a minimum of 1000 cards.

Plant and equipment

The list of available equipment is presented in Table 3.2.

Inventory

The inventory of the company corresponds to the list of equipment. Therefore, we included a new column in Table 3.2 with the corresponding total cost.

Table 3.2 List of equipment

Equipment	Purchase Date	Purchase Price (€)	Running Cost (€)	Total Cost (€)
Portable laptop (3x)	15/01/2017	750	100	2250
GPS Device (2x)	15/01/2017	200	60	400
Software (3x)	30/01/2017	300	100	900
Smartphones (3x)	01/01/2017	250	600	750
Printer	30/01/2017	200	150	200
Server	01/01/2017	1500	300	1500
Tablets (2x)	15/03/2017	300	100	600

Software

The company needs two kinds of software applications:

- Website – includes the cost of hosting per year. This cost is around 100 € per year;
- Accounting package – Sage software will be adopted as accounting software. This cost is around 200 € per year.

The total cost of software is around 300 €, as stated in Table 3.2.

Communication channels

The company will interact with its customers using social networks, email, and telephone. The headquarters of the company will not offer a direct contact point with customers. Besides that, the company communicated with tourism agencies using the same communication channels and the commercial team.

Payments types accepted

The company directly only accepts MB NET and credit card as form of payment. However, tourism agencies can offer other forms of payments including cash.

Credit policy

The company uses two different forms of payment policies. For end-users, the prepayment model is adopted. For tourism agencies, a 30-day payment policy is used. Finally, for payment to suppliers, we also use a 30-day payment policy.

Quality control

The company adopts a strict control over the choice of its partners and we will act, in legal compliance, to safeguard our partnerships. The relationship with partners will be evaluated annually. With regard to our clients, an assessment of the degree of satisfaction will be made, immediately after the date of the service rendering, in order to evaluate the quality of the service. This evaluation will be done through online or telephone surveys, according to customer preference and availability.

3.2.2 Action Plan

Table 3.3 presents several milestones considered in our action plan. The elements are ordered considering the date of expected completion.

Table 3.3 Planned tasks for the action plan

Milestone	Date of Expected Completion	Person/Role Responsible
Registration of the company	15/01/2017	General manager
Market research through interviews and surveys	15/02/2017	Marketing
Webpage development	28/02/2017	IT
Development of social marketing (e.g., Facebook, Pinterest, Twitter, YouTube, etc.)	31/03/2017	Marketing
Launch of the company in the city of Porto	15/04/2017	General manager and marketing
Partner recruitment	31/05/2017	General manager and marketing
Carry out the training plan	31/05/2017	General manager
Implementation of marketing strategy and advertising in buses and airports	30/09/2017	Marketing
Launch of the company in the city of Lisbon	31/10/2017	General manager and marketing
Implementation of marketing strategy and advertising in AdBikes	30/11/2017	Marketing
Development of a new service: online tour guide	31/03/2018	IT and marketing
Launch of the application for smartphones	31/03/2018	IT
Expansion for Portuguese-speaking countries (preferably Brazil and Angola)	30/06/2018	General manager and marketing
Expansion to Foreign Language & Cross-Platform Countries (the United States and England)	30/06/2019	IT and customer support service

3.3 Scenario II – AuditExpert

3.3.1 Operations

Production process

The company uses the waterfall software methodology. The waterfall or cascade model is an engineering model designed to be applied in software development. The main idea behind it is that the different stages of development follow a sequence. In this way, the output of the first stage flows to the second stage, the output of the second flows to third, and so on. The activities to be performed are grouped into tasks, executed sequentially, so that a task can only start when the previous task has finished.

The waterfall software model has the advantage that it only advances to the next task when the client validates and accepts the end products of the current task. In order to decrease the rigidity of the model and increase the level of participation of our clients, we use a slightly different notion of the cascade model, whose main difference is to predict the possibility of any tasks in the cycle being able to return to a previous task. This change is particularly suitable to contemplate the introduction of new functional/technical changes that have arisen in the meantime. On the contrary, the main risk of this approach is that, in the absence of a well-defined project management and change control process, we can spend time in an infinite cycle, without ever reaching the end goal. In order to face this challenge, we establish that each cycle cannot be above 1/2 of the estimated total duration of the project.

The proposed software methodology is composed of five stages (Figure 3.2):

- Requirements analysis and definition – in the initial phase of the model, we look into the functional and non-functional requirements of the system, which usually consists of the services that must be provided, as well as limitations and objectives of the software. This initial step also includes documentation (e.g., UML use cases specification) and study of the ease and feasibility of the project (e.g., risk analysis) in order to determine the process of beginning the development of the system project;
- Design – the software design process is a multi-step process that focuses on four different system attributes: data structure, software architecture, procedural details, and interface characterization. The design process represents the requirements in a way that allows product coding;
- Implementation – this is the stage at which programs are created. If the project has a high level of detail, the coding step can be implemented using automatic software generation. This process creates the class of the systems, methods, attributes, and associated unit tests;
- Verification – once the coding step is completed, the system test phase can begin. The testing process focuses on the main points: (i) the internal software logic and (ii) the external functionalities. This phase decides whether software "behavior" errors have been solved and ensures that the inputs defined produce current results that match the specified requirements. The test phase shall also include all the acceptance tests relative to each functional and non-functional requirement;

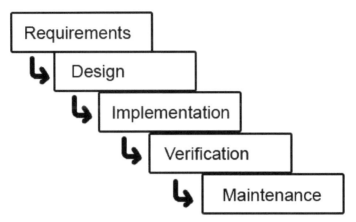

Figure 3.2   Waterfall methodology.

- Maintenance – this step consists in correcting errors that were not previously detected, in functional and preference improvements and other types of support.

Suppliers

The company has the following suppliers:

- OpenVAS – provides a framework composed by several services and tools for vulnerability management and scanning. These tools are integrated in the communication security package;
- ZoneMinder – provides tools for monitoring the IT infrastructure. These tools are integrated in the platform and infrastructure package.

Plant and equipment

The list of available equipment is presented in Table 3.4.

Inventory

The inventory of the company corresponds to the list of equipment. Therefore, we included a new column in Table 3.4 with the corresponding total cost.

Software

The company uses four types of software applications:

- Website – includes the cost of hosting per year. This cost is around 200 € per year;

Table 3.4 List of equipment

Equipment	Purchase Date	Purchase Price (€)	Running Cost (€)	Total Cost (€)
Portable laptop (5x)	9/01/2017	1000	300	5000
Software (3x)	15/01/2017	350	400	350
Smarphones (3x)	9/01/2017	300	300	900
Printer (2x)	30/01/2017	150+250	400	400
Server (2x)	30/01/2017	2500	600	5000
IP Cams (2x)	30/01/2017	150	50	300

- Accounting package – PHC software will be adopted as accounting software. This cost is around 150 € per year;
- Network tool – Open-AudIT is used to draw the network, shows how it is configured, and gives alerts when it changes. It is an open-source software with a GPL license;
- Auditor package – Eramba is used to manage, analyze, and report the security governance. It is an open-source software with the MIT license.

The total cost of software is around 350 € as stated in Table 3.4. However, the running costs of software are significant (around 400 € per year) due to the need of internal support for configuration and updates management of Open-AudIT and Eramba.

Communication channels

In this market, the establishment of a communication policy very close to the various partners and customers is vital. For that, the use of social networks and online chats in the website of the company will play an important role. Besides that, we will frequently be present in technical magazines and we will have a regular participation in international fairs. Finally, we will use endorsement campaigns that will help us to reach some important business clients and will improve our credibility in the market.

Payments types accepted

The company accepts payments by cash, bank transfer, paypal, bank checks, credit cards, debit cards, and courier vouchers. The customer can freely choose the type of payment without any penalty in the final price of the product/service.

Credit policy

We adopt a policy of a maximum of 60 days for payments to our suppliers and receipts from our clients.

Quality control

The company uses the following certifications:

- ISO/IEC 27001 Information security management – helps our company to manage the security of assets in the context of an Information Security Management System (ISMS), which is a systematic approach to managing sensitive company information so that it remains secure. It includes people, process, and IT systems by applying a risk management process;
- ISO 9001:2008 Quality management systems – specifies requirements for a quality management system. This standard lets us to demonstrate our ability to consistently provide products and services that meets customer and applicable statutory and regulatory requirements. Furthermore, it aims to enhance continual improvement of the system and the assurance of conformity to customer and applicable statutory and regulatory requirements.

Table 3.5   Planned tasks for the action plan

Milestone	Date of Expected Completion	Person/Role Responsible
Registration of the company	31/01/2017	General manager
Market research analysis	31/01/2017	General manager
Bank loan	31/01/2017	General manager
Agreement with risk capital entity	31/01/2017	General manager
Partnership agreements	15/02/2017	General manager and consultants
Webpage development	28/02/2017	Consultants
Presence in social networks (e.g., LinkedIn, Facebook, YouTube, and Twitter)	28/02/2017	General manager
Identification of target technical magazines	28/02/2017	General manager
Launch of the company in the Portuguese market	31/03/2017	General manager
At least two participations in international technical fairs and job fairs	30/06/2017	General manager
Availability of audit module	30/06/2017	Consultants
Endorsement campaign is defined	30/06/2017	General manager
Own installations	31/01/2019	General manager
Entrance in the UK market	31/03/2019	General manager
Entrance in the Latin America market	31/05/2020	General manager

These two certifications are very relevant and mandatory for a considerable amount of our partners and clients. In order to guarantee its adoption, we perform two audits per year of ISO/IEC 27001 and one audit per year for ISO 9001:2008.

3.3.2 Action Plan

The action plan of our company is composed by the list of activities presented in Table 3.5. These activities are ordered by date of expected completion.

4

Financial Plan & Viability Analysis

Overview

The financial plan and viability analysis must provide an evaluation and correction of the company's direction. In addition, it should reflect the financial impact of the various elements presented in the other sections of the business plan, such as marketing expenses, personnel expenses, fixed costs, sales projection, and profitability analysis considering a set of indicators. We start by defining the financial objectives of the company, identifying the necessary financing and carrying out a cash flow forecast. Finally, the analysis of financial profitability takes into account a 6-year projection, considering three scenarios (standards, pessimistic, and optimistic). The changes considered in each of these scenarios are highlighted. Finally, we present a financial and viability analysis applied to *TourMCard* and *AuditExpert*.

4.1 Theoretical Foundations

4.1.1 Key Objectives and Financial Review

Financial planning is the process of estimating the capital required for the constitution and operation of the business. Financial planning has two strategic objectives:

1. Determining capital requirements – this will depend upon factors like cost of current and fixed assets, promotional expenses, and long-range planning. Capital requirements should to be looked from the short-term and long-term perspectives. In the IT field, it is common to find small amount of capital requirements, but there is a great need for reinvestments throughout the company's activity;

2. Determining the capital structure – the capital structure is the composition of capital, if it belongs to the promoters of the business or if it requires some source of external aid, such as banks, micro credit, or venture capital.

Financial planning is very important to the promoters and potential investors. Its importance can be outlined in the following dimensions:

- Adequate funds have to be ensured;
- Financial planning helps in ensuring a reasonable balance between outflow and inflow of funds so that stability is maintained;
- Financial planning ensures that the suppliers of funds are easily investing in companies that exercise financial planning;
- Financial planning helps in establishing growth and expansion programs, which helps in the long-run survival of the company;
- Financial planning reduces uncertainties with regard to changing market trends, which can be faced easily through enough funds;
- Financial planning helps in reducing the uncertainties, which can be a hindrance to the growth of the company.

One of the initial steps when creating a financial plan is setting goals and objectives. Therefore, business promoters can set different types of objectives, including financial objectives that will give them a solid plan for moving in the direction of long-term success. Common financial business objectives include increasing revenue, increasing profit margins, sustainability, or return on investment.

Another important issue is to determine start-up costs. These costs represent one-time expenditures that the start-up company must make before it opens its doors for business. Such costs typically include equipment, furniture, fixtures, suppliers and materials, inventory, renovations (leasehold improvements), licenses, permits, and incorporation fees (if applicable).

After this initial phase, we can start working on financial statements. They typically come in three documents: income statement, balance sheet, and cash flow statement. Together they provide an accurate picture of a company's current value and its ability to face financial duties along the time.

The income statement, also referred to as income, profit and loss statement (P&L) illustrates the profitability of a company over a given period of time. This statement typically includes one section detailing revenues and gains and another section detailing expenses and losses. The analysis of an income statement is very easy: if the company's revenues and gains are greater than its expenses and losses, then the income statement will show a net profit; on the other hand, if the company experiences greater expenses and losses, then the income statement shows a net loss.

The balance sheet provides an overview of the company's financial situation at a specific time, rather than profitability over a period of time. The balance sheet includes the company's assets, liabilities, and owner's equity.

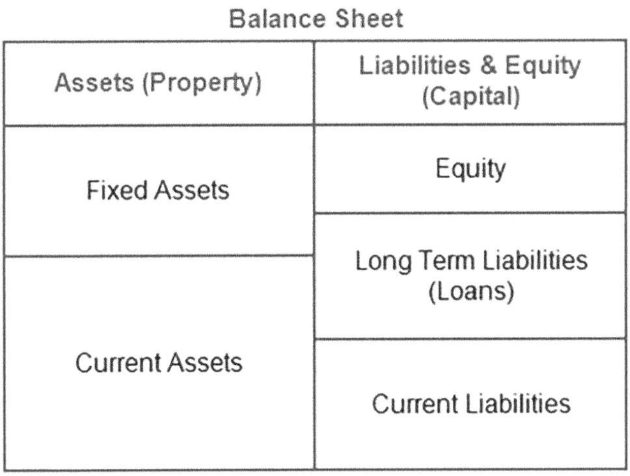

Balance Sheet

Assets (Property)	Liabilities & Equity (Capital)
Fixed Assets	Equity
	Long Term Liabilities (Loans)
Current Assets	
	Current Liabilities

Figure 4.1 Structure of a balance sheet.

The company's assets must always equal its liabilities plus owner equity. A picture of the balance sheet structure is given in Figure 4.1.

The cash flow statement records the amount of actual money that flows into and out of the company. It allows investors to understand how a company's operations are running, where its money is coming from, and how it is being spent. The cash flow is determined by looking at three components by which cash enters and leaves a company: core operations, investing, and financing.

There is a typical misunderstanding between the concepts of income statement and cash flow statement. The major difference between these two statements is that the income statement is based on an accrual basis (due or received) while the cash flow statement is based on the actual receipt and payment of cash. Furthermore, the income statement is classified into two main activities: operating and non-operating, whereas the cash flow statement is divided into three activities: operating, investing, and financing. Additionally, the income statement is helpful in knowing the profitability of the company, but the cash flow statement is useful in knowing the present and future cash flows.

4.1.2 Viability Analysis

There are three key financial indicators that can be used to study the viability of a business. These elements are: net present value (NPV), internal rate of return (IRR), and payback.

The NPV is represented as the mathematical-financial formula capable of determining the present value of future payments discounted at an appropriate interest rate, less the cost of the initial investment. The formula can be expressed as follows, where **t** represents the cash flow period and **i** is the discount rate:

$$NPV = -Initial\ Investment \sum_{t=1}^{T} \frac{Net\ Cash\ Flow_t}{(1+i)^t} \qquad (4.1)$$

In order to properly understand all the elements of the formula (4.1), we need also to define the following concepts:

- Present value – the current worth of a future sum of money or stream of cash flows given a specified rate of return. Future cash flows are discounted at the discount rate, and higher the discount rate, lower the present value of the future cash flows;
- Annuity – a series of equal payments or receipts that occur at evenly spaced intervals, such as leans, rental payments, regular deposits, etc.;
- Perpetuity – a constant stream of identical cash flow with no end. The perpetuity can be calculated by the sum of estimated cash flows divided by a given or pre-determined discount rate. The perpetuity value should be estimated fairly moderately in IT business, considering the high existence competition, and the constant emergence of new technological solutions and new ventures.

The present value appears always in the process of calculating the NPV. The other two elements (annuity and perpetuity) do not appear always in all business scenarios, but it is relevant to know and use them when needed.

Figure 4.2 presents a comparative scenario between two investments. Project Y offers a higher NPV for a rate of discount below 10.9%. On the other hand, project Z has a higher NPV for rate of discounts higher than 10.9%. Finally, the NPV is equal to both projects when the rate of discount is equal to 10.9%.

The IRR is a mathematical-financial formula used to calculate the discount rate that would have a certain cash flow to equal its net present value at zero. The formula can be expressed as follows:

$$\sum_{t=0}^{T} \frac{Net\ Cash\ Flow_t}{(1+IRR)^t} = 0 \qquad (4.2)$$

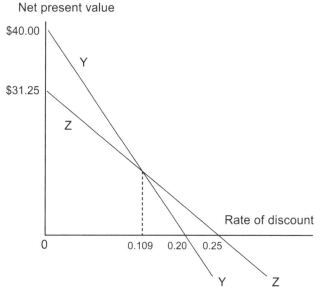

Figure 4.2 Analysis of NPV between two projects.

Due to the nature of formula (4.2), IRR cannot be calculated analytically and must instead be calculated either through trial-and-error methods or by using software programs like MS Excel.

The IRR should be used in conjunction with NPV for a clearer picture of the value represented by a potential investment. If we use it separately, two kinds of issues may arise. First, depending on the initial investment costs, a project may have a low IRR, but a high NPV, meaning that while the pace at which the company sees returns on that project may be slow, the project may also be adding a great deal of overall value to the company. Second, a similar issue appears when using the IRR to compare projects of different lengths. For example, a project of a short duration may have a high IRR, making it appear as an excellent investment, but may also have a low NPV. Conversely, a longer project may have a low IRR, earning returns slowly and steadily, but may add a large amount of value to the company over time.

The payback represents the amount of time needed to recover the initial investment made on the project. This period is not always short, but it depends on the value of the investment and the type of business. In general, the return may happen from months to years. In the IT field, we have typically associated very short paybacks, in some cases even less than a year of activity.

The main advantages associated with the payback indicator are: (i) it presents simple formula, easy to be applied and learned; (ii) it provides an idea of the level of liquidity of the business and the level of risk involved; (iii) it is especially useful in projects with very a high risk degree or in projects with limited life; and (iv) in times of financial crisis and economic instability, the resource serves to increase business security. On the other hand, the main disadvantages are: (i) the indicator values differently the flows received in different periods (this is in accordance with the dualistic thought, before or after the payback, disregarding the values received within each of these intervals); (ii) for longer duration projects, the feature is not very recommended as it does not consider the cash flows produced after the year of recovery.

A slightly different notion of payback, called discounted payback, appeared in the literature in order to increase the robustness of pay-back calculus. The discounted payback period reflects the amount of time necessary to break even in a project based not only on what cash flows occur, but also on when they occur and the prevailing rate of return on the market. These two calculations, although similar, may not return the same result due to the discounting of cash flows. The discounted payback offers the following advantages: (i) it remains simple and practical, like a simple payback and (ii) it solves the problem of not considering the value of money in time. On the other hand, despite considering a discount rate, it continues without taking into account cash flows after the payback period.

Another perspective to perform a viability analysis is to look into the critical elements in estimated cash flows that can significantly change the profitability of a business. A sensitivity analysis is a technique used to determine how different values of an independent variable impact a particular dependent variable under a given set of assumptions. In order to perform a sensitivity analysis, we should do it in a systematic and consistent manner. Therefore, the following steps should be carried out: (i) identification of the critical variables which affect the project decision; (ii) calculation of the effects of these changes; and (iii) consideration of the variables in possible combinations that can be changed simultaneously.

Another instrument to analyze the viability of an investment or start-up business is performing a business ratio analysis. There are a significant number of business ratios which give us a way to assess the relative health of a business. Ratio analysis is a useful management tool because it helps identify positive and negative trends in the business performance. The data for the ratio analysis come from the balance sheet, income statement, and cash flow statement. The ratios should be compared to the ratios of similar businesses, which can be obtained from industry associations, government entities, and business libraries.

There are six aspects of operating performance and financial condition, and we can evaluate from financial ratios:

- Liquidity ratio – provides information on a company's ability to meet its short-term obligations;
- Profitability ratio – provides information on the amount of income from each sale;
- Activity ratio – relates information on a company's ability to manage its resources efficiently;
- Financial leverage ratio – provides information on the degree of a company's fixed financing obligations and its ability to satisfy these financial obligations;
- Shareholder ratio – describes the company's financial condition in terms of amounts per share of stock;
- Return on investment ratio – provides information on the amount of profit, relative to the assets employed to produce that profit.

Liquidity ratios provide a measure of a company's ability to generate cash to meet its immediate needs. Two of the most important liquidity ratios are:

- Current ratio – the ratio of current assets to current liabilities. It indicates a company's ability to satisfy its current liabilities with its current assets:

$$Current\ ratio = \frac{Current\ assets}{Current\ liabilites} \tag{4.3}$$

- Quick ratio – the ratio of quick assets to current liabilities. It indicates a company's ability to satisfy current liabilities with its most liquid assets.

$$Quick\ ratio = \frac{Current\ assets - Inventory}{Current\ liabilites} \tag{4.4}$$

Profitability ratios compare components of income with sales. They give us an idea of what makes up a company's income. There are three common profitability ratios:

- Gross profit margin – the ratio of gross income or profit to sales. It indicates how much of the money from sales is left after costs of goods sold:

$$Gross\ profit\ margin = \frac{Gross\ income}{Sales} \tag{4.5}$$

- Operating profit margin – the ratio of operating profit (EBIT) to sales. It indicates how much money is left over after operating expenses:

$$Operating\ profit\ margin = \frac{Operating\ income}{Sales} \tag{4.6}$$

- Net profit margin – the ratio of net income to sales. It indicates how much money of sales is left over after all expenses.

$$Net\ profit\ margin = \frac{Net\ income}{Sales} \tag{4.7}$$

Activity ratios are measures of how well assets are used. They can be used to evaluate the benefits produced by all company's assets collectively, or specific assets, such as inventory or accounts receivable. These measures give us an idea of how effectively the company is at putting its investment to work. The greater the turnover, the more effectively the company is at producing a benefit from its investment in assets. The two most common turnover ratios are the following:

- Inventory turnover – the ratio of cost of goods sold to inventory. It indicates how many time inventory is created and sold during the period:

$$Inventory\ turnover = \frac{Cost\ of\ goods\ sold}{Inventory} \tag{4.8}$$

- Total asset turnover – the ratio of sales to total assets. It indicates the extent that the investment in total assets results in sales:

$$Total\ asset\ turnover = \frac{Sales}{Total\ assets} \tag{4.9}$$

Financial leverage ratios are used to assess how much financial risk the company has taken on. The two most common ratios in this domain are:

- Total debt to assets ratio – it indicates the proportion of assets that are financed with debt (both short-term and long-term debts):

$$Total\ debt\ to\ assets\ ratio = \frac{Total\ debt}{Total\ assets} \tag{4.10}$$

- Long-term debt to assets ratio – it indicates the proportion of the company's assets that are financed with long-term debt:

$$Long-term\ debt\ to\ assets\ ratio = \frac{Long-term\ debt}{Total\ assets} \tag{4.11}$$

Shareholders ratios provide information for managers and creditors about the financial dividends paid by the company. The most knowledge ratio is the following:

- Earnings per share (EPS) – the amount of income earned during a period per share of common stock:

$$Earnings\ per\ share = \frac{Net\ income\ available\ to\ shareholders}{Number\ of\ shares\ outstanding} \quad (4.12)$$

- Return on equity (ROE) – it measures the rate of return that shareholders receive on their investment in the business. In fact, it will inform shareholders how much the company is earning compared it to the performed investment:

$$ROE = \frac{Net\ income\ for\ the\ year - taxes - interest}{Shareholders'\ equity} \quad (4.13)$$

- Return on investment ratio is a very popular metric because of its versatility and simplicity. It is used to evaluate the efficiency of an investment or to compare the efficiency of a number of different investments. To calculate this ratio, we need to use the expression below:

$$ROI = \frac{Gain\ from\ investment - cost\ of\ investment}{Cost\ of\ investment} \quad (4.14)$$

Finally, it is important to consider a scenario analysis in order to evaluate the risk associated with the expected cash flows. Typically, our projected cash flows are estimated under the most likely scenario. However, it is important to estimate the expected cash flow and asset value under various scenarios, with the intent of getting a better sense of the effect or risk on the value.

A scenario can be defined as a possible future environment, either at a point in time or over a period of time. A projection of the effects of a scenario over the time period studies should incorporate several risk factors, often over multiple time periods. Scenarios can also be complex, involving changes to and interactions among many factors over time, perhaps generating a set of cascading events. Because the future is uncertain, there are many possible scenarios and, in addition, there may be a range of financial effects on a form arising from each scenario.

The most basic approach is to consider three scenarios (the normal expected cash flow, the best case, and the worst case). In practice, there are two ways in which this analysis can be structured. In the first, each input into asset value is set to its best (or worst) possible outcome and the cash flows estimated with those values. The problem with this approach is that it may not be feasible; after all, to get higher revenue growth, the firm may have to lower prices and accept lower margins. In the second, the best possible scenario is defined in terms of what is feasible while allowing the relationship between the inputs. Thus, instead of assuming that revenue growth and margins will both be maximized, we will choose that combination of growth and margin that is feasible and yields the best outcome. While this approach is more realistic, it does require more work to put into practice.

Finally, we can adopt a multiple scenario analysis approach. In fact, scenario analysis does not have to be restricted to the best and worst cases. In its most general form, the value of a risky asset can be computed under a number of different scenarios, varying the assumptions about both macroeconomic and asset-specific variables.

4.2 Scenario I – TourMCard

4.2.1 Key Objectives and Financial Review

Our financial goal is to have a low payback time using a small amount of investment. We expect to have a payback under the first 2 years of activity. We will not get any financial aid from banks or other kinds of financial entities.

A first step is to calculate the start-up costs, which are depicted in Figure 4.3. The total start-up costs are estimated around 23,000 €. In our business, we considered the following structure of costs:

- Leasehold improvements – in this section, we include the renting of installations in the science park;
- Equipment – the existence of two laptops;
- Furniture – office supplies that could be needed;
- Pre-opening salaries and wages – we considered a gross salary of 1200 € for two persons during 3 months. A gross salary of 1200 €, attending to Portuguese final taxes, will involve a total cost of 2100 € per month for each employee. After this initial period, we considered that the revenues will be enough to pay it;
- Supplies – we plan to have a delivery of 1000 smart cards for the period of three months. The cost of each smart card ia around 2 €;
- Advertising and promotions – we estimate an initial cost of 2500 € in advertising, which includes social network presence, email marketing, and advertising on buses and ad-bike (two times per month);
- Licenses – a cost of 100 € for the IOS development license;
- Working capital – an estimated value of 2500 € as working capital.

Then, we analyze the sources of funding in Figure 4.4. The entire investment is exclusively made by the business' promoters.

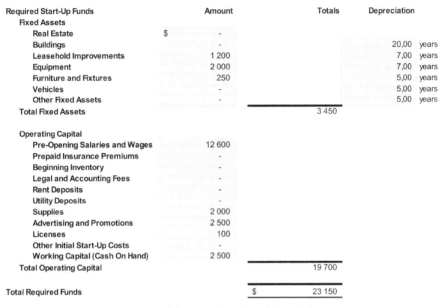

Required Start-Up Funds	Amount	Totals	Depreciation	
Fixed Assets				
Real Estate	$ -			
Buildings	-		20,00	years
Leasehold Improvements	1 200		7,00	years
Equipment	2 000		7,00	years
Furniture and Fixtures	250		5,00	years
Vehicles	-		5,00	years
Other Fixed Assets	-		5,00	years
Total Fixed Assets		3 450		
Operating Capital				
Pre-Opening Salaries and Wages	12 600			
Prepaid Insurance Premiums	-			
Beginning Inventory	-			
Legal and Accounting Fees	-			
Rent Deposits	-			
Utility Deposits	-			
Supplies	2 000			
Advertising and Promotions	2 500			
Licenses	100			
Other Initial Start-Up Costs	-			
Working Capital (Cash On Hand)	2 500			
Total Operating Capital		19 700		
Total Required Funds		$ 23 150		

Figure 4.3   Required financial capital.

Sources of Funding	Amount	Totals	Loan Rate	Term in Months	Monthly Payments
Owner's Cash Injection	100,00%	23 150			
Outside Investors	0,00%	-			
Additional Loans or Debt					
Commercial Loan	0,00%	-	9,00%	84,00	$0,00
Commercial Mortgage	0,00%	-	9,00%	240,00	$0,00
Total Sources of Funding	100,00%	$ 23 150			$0,00

Figure 4.4   Sources of funding.

The projected cash flows are given in Figure 4.5. The analysis period considered is 2017–2022, where the cash flows are estimated during 6 years. In incomes section, we include earnings from product and services sales from national and international markets; in cost section, we include rubrics such as salaries, advertising, and equipment. In more detail, the following elements were considered as incomes:

- Product sales from national market – in 2017, we predict sales of 200 smart cards per month, and 500 cards for 2018 per month with an income of 5 € for each sale. After 2019, we establish an increase of 10% in revenue per year;
- Product sales from international market – revenues from international markets will only appear in 2018. In this year, we estimate sales of 1000 smart cards with an increase of 20% in revenues for next years;

	2017	2018	2019	2020	2021	2022
Incomes						
Product Sales (National market)	12 000,00€	30 000,00€	33 000,00€	36 300,00€	39 930,00€	43 923,00€
Product Sales (International market)	0,00€	60 000,00€	72 000,00€	86 400,00€	103 680,00€	124 416,00€
Services (National market)	24 000,00€	120 000,00€	132 000,00€	145 200,00€	159 720,00€	175 692,00€
Services (Internationl market)	0,00€	120 000,00€	150 000,00€	187 500,00€	234 375,00€	292 968,75€
Other incomes	2 750,00€	3 437,50€	4 296,88€	5 371,09€	6 713,87€	8 392,33€
Total of incomes	38 750,00€	333 437,50€	391 296,88€	460 771,09€	544 418,87€	645 392,08€
Costs						
Investments	23 150,00€	0,00€	0,00€	0,00€	0,00€	0,00€
Loan Payments	0,00€	0,00€	0,00€	0,00€	0,00€	0,00€
Buildings +Renting	0,00€	1 200,00€	18 000,00€	18 000,00€	18 000,00€	18 000,00€
Salaries	56 700,00€	126 000,00€	216 000,00€	270 000,00€	360 000,00€	360 000,00€
Equipments	1 000,00€	2 000,00€	3 000,00€	2 000,00€	2 000,00€	0,00€
Advertising	10 000,00€	30 000,00€	10 000,00€	10 000,00€	10 000,00€	10 000,00€
Supplies	2 800,00€	24 000,00€	42 000,00€	49 080,00€	57 444,00€	67 335,60€
Travel	22 000,00€	27 000,00€	17 000,00€	17 000,00€	17 000,00€	17 000,00€
Vehicles	0,00€	0,00€	0,00€	0,00€	0,00€	0,00€
Telephone and Communications	2 400,00€	3 600,00€	3 780,00€	3 969,00€	4 167,45€	4 375,82€
Miscellaneous services	2 000,00€	2 100,00€	5 000,00€	5 250,00€	5 512,50€	5 788,13€
Total of costs	120 050,00€	215 900,00€	314 780,00€	375 299,00€	474 123,95€	482 499,55€
Net Income	-81 300,00€	117 537,50€	76 516,88€	85 472,09€	70 294,92€	162 892,54€
Discounted net income	-81 300,00€	102 206,52€	57 857,75€	56 199,29€	40 191,35€	80 986,38€

Figure 4.5   Projected cash flows for standard scenario.

- Services' revenues from national market – in 2017, we estimate that consumers will use 200 packs per year, and 1000 packs for 2018. After that, we estimate an increase of 10% per year in the revenue;
- Services from international market – in 2018, the overall services in the internal market will be approximately 120,000 € (1000 packs sales). In 2018, we estimate that 50% of revenue will come already from the international market. After that, we estimate an increase of 25% per year of our services in the international market. For 2022, the revenues from international markets (products + services) will represent 65% of our total revenue;
- Other incomes – in this section, we include the gift voucher cells. We estimate to have 50 sells per year of vouchers around 10 €, also 50 sells for vouchers with an average price of 25 €, and 10 sells for vouchers with a maximum price of 100 €.

Now we can also look for our cost structure:

- Investments – we only consider an initial investment of 23,150 €. Other investments needed along the time will be placed in other sections, such as advertising, equipment, or travel;
- Buildings + Renting – for 2017, we predict to use our initial investment to pay these expenses. For 2018, we estimate a total cost of 1200 € per year. Then, in 2019, we will intend to have our own facilities with an estimated rent of 18,000 € per year;

- Salaries – for 2017, we estimate to have three employees in our company with an average cost of 2100 € per month. For 2018, we estimate to have five employees. For 2019, the average cost of each employee will increase to 2250 € and we expect to have a total of eight employees. For 2020, the total number of employees will be 10. For 2021 and 2022, we expect to have 12 employees with an average cost of 2500 €. Making a brief analysis, we can conclude that in 2017 around 47% of our expenses will come from salaries and this percentage will increase to around 75% for 2022;
- Equipment – in this section, we include only the costs of laptops. We expect to have a cost of 1000 € for each employee. In 2017, we include a cost of one laptop, because in the initial investment, we already had included two laptops;
- Advertising – for 2017, we expect to have a total cost of 10,000 € in advertising. This amount will increase significantly to 30,000 € in 2018, due to our presence in international markets. Then, in 2019, our investments in advertising will be reduced to 10,000 € and will be stabilized at this level for the next years;
- Supplies – the cost of supplies is estimated based on our sales of smart cards in national and international markets. The cost of our sales represents 40% (2 € in 5 €);
- Travel – in this section, we include costs from our domestic and foreign trips. In our domestic component, we estimate an average cost of 1000 € per month; for foreign markets, we estimate four travels for 2017 (trip + registration + other costs) of 2500 € per each presence in internal fairs. The presence in international fairs will increase to 6 in 2018, will decrease to 2 in 2019, and will stabilize at this value for further years;
- Telephone and communications – for 2017, we estimate an average cost of 200 € per month. This value will be increased to 300 € per month for 2018. For next years, we estimate an increase of 5%;
- Miscellaneous services – other types of costs associated with services have an estimation of 2000 €. This amount will increase to 5% for next years.

4.2.2 Viability Analysis

In order to perform a viability analysis, we considered three scenarios: standard, pessimist, and optimist. We detail the different inputs and characteristics of each scenario and we calculate the NPV, IRR, and payback. We considered a discount rate of 15% in all scenarios.

Standard Scenario

The standard scenario uses the same cash flow projections available in Figure 4.5. The values of NPV, IRR, and payback are available in Table 4.1. The analysis of the most relevant ratios is given in Table 4.2.

Our analysis indicates an NPV higher than 250,000 € for 6 years and an IRR of 123.50%. NPV higher than 0 and IRR higher than the discount rate of 15% indicate that the project is profitable, besides the payback is 1 year and 8 months.

The most relevant considered ratios in our business are the gross profit margin (GPM), return on equity (ROE), and return on investment (ROI). The GPM remains at 60% in the entire period, because the unit sales price (5 €) and the unit cost of product sales (2 €) are always same for the 6 years. The ROE changes according to the evolution of net income. It is important to highlight that we considered a tax of 23% plus 5% as interest rate. Finally, the ROI represents the gain of an investment for each year. The ROI reaches its maximum value for 2018.

Pessimistic Scenario

The projected cash flows are given in Figure 4.6. The following elements were introduced in the pessimistic scenario:

- The products sales and services were reduced by 25% for all considered years;
- Costs of advertising increase 10% in the first 2 years. We considered also that the costs of advertising will remain the same for the next years;
- Costs of supplies increase 10% in all considered period.

The pessimistic scenario will have a strong impact on our business, particularly after 2018, if corrective measures are not taken in order to adjust our

Table 4.1 Financial indicators for the standard scenario

NPV	256,141.29 €
IRR	123.50%
Payback	1.69 (1 year and 8 months)

Table 4.2 Ratios for the standard scenario

Year	2017	2018	2019	2020	2021	2022
GPM	60%	60%	60%	60%	60%	60%
ROE	−351.19%	365.56%	237.98%	265.83%	218.63%	506.62%
ROI	−67.72%	54.44%	24.31%	22.77%	14.83%	33.76%

	2017	2018	2019	2020	2021	2022
Incomes						
Product Sales (National market)	9 000,00€	22 500,00€	18 562,50€	15 314,06€	12 634,10€	10 423,13€
Product Sales (International market)	0,00€	45 000,00€	40 500,00€	36 450,00€	32 805,00€	29 524,50€
Services (National market)	18 000,00€	90 000,00€	74 250,00€	61 256,25€	50 536,41€	41 692,54€
Services (Internationl market)	0,00€	90 000,00€	84 375,00€	79 101,56€	74 157,71€	69 522,86€
Other incomes	2 750,00€	3 437,50€	4 296,88€	5 371,09€	6 713,87€	8 392,33€
Total of incomes	29 750,00€	250 937,50€	221 984,38€	197 492,97€	176 847,09€	159 555,36€
Costs						
Investments	23 150,00€	0,00€	0,00€	0,00€	0,00€	0,00€
Loan Payments	0,00€	0,00€	0,00€	0,00€	0,00€	0,00€
Buildings +Renting	0,00€	1 200,00€	18 000,00€	18 000,00€	18 000,00€	18 000,00€
Salaries	56 700,00€	126 000,00€	216 000,00€	270 000,00€	360 000,00€	360 000,00€
Equipments	1 000,00€	2 000,00€	3 000,00€	2 000,00€	2 000,00€	0,00€
Advertising	11 000,00€	33 000,00€	10 000,00€	10 000,00€	10 000,00€	10 000,00€
Supplies	1 760,00€	26 400,00€	25 987,50€	22 776,19€	19 993,20€	17 576,96€
Travel	22 000,00€	27 000,00€	17 000,00€	17 000,00€	17 000,00€	17 000,00€
Vehicles	0,00€	0,00€	0,00€	0,00€	0,00€	0,00€
Telephone and Communications	2 400,00€	3 600,00€	3 780,00€	3 969,00€	4 167,45€	4 375,82€
Miscellaneous services	2 000,00€	2 100,00€	5 000,00€	5 250,00€	5 512,50€	5 788,13€
Total of costs	120 010,00€	221 300,00€	298 767,50€	348 995,19€	436 673,15€	432 740,91€
Net Income	-90 260,00€	29 637,50€	-76 783,13€	-151 502,22€	-259 826,06€	-273 185,55€
Discounted net income	-90 260,00€	25 771,74€	-58 059,07€	-99 615,17€	-148 556,40€	-135 821,50€

Figure 4.6 Projected cash flows for pessimistic scenario.

salary structure that is our main cost. Only NPV will only be positive for 2018. The value of NPV, IRR, and payback is available in Table 4.3. The analysis of the most relevant ratios is given in Table 4.4.

Our analysis indicates a negative NPV, which means that our business is not profitable. In such situation, the IRR and payback cannot be calculated.

The GPM remains at 60% in the entire period, because we did not change the unit price of sales or purchases. The ROE is very volatile to the high level of variety of net income that becomes strongly negative for 2019 and next years. Finally, the ROI is always negative, except for 2018.

Table 4.3 Financial indicators for the pessimistic scenario

NPV	−506,540.40 €
IRR	Not applied
Payback	Not applied

Table 4.4 Ratios for the pessimistic scenario

Year	2017	2018	2019	2020	2021	2022
GPM	60%	60%	60%	60%	60%	60%
ROE	−389.89%	92.18%	−331.68%	−654.44%	−1122.36%	−1180.07%
ROI	−75.21%	13.39%	−25.70%	−43.41%	−59.50%	−63.13%

Optimistic Scenario

The projected cash flows are given in Figure 4.7. The optimistic scenario introduces the following changes compared to the standard scenario:

- The products sales and services were increased to 25% for all considered years;
- Costs of advertising increase 20% in the considered period. It is expected that the increase in sales also originates an increased advertising spending.

The optimistic scenario will also have a strong impact on our business, particularly in the last 2 years. The values of NPV, IRR, and payback are available in Table 4.5. The analysis of the most relevant ratios is given in Table 4.6.

The NPV increases significantly to more than 1700 m€. The IRR almost triplicates considered the standard scenario. The payback is reduced in around 3 months, which is not particularly significant considering the impact in terms of NPV and IRR.

	2017	2018	2019	2020	2021	2022
Incomes						
Product Sales (National market)	15 000,00 €	37 500,00 €	51 562,50 €	70 898,44 €	97 485,35 €	134 042,36 €
Product Sales (International market)	0,00 €	75 000,00 €	112 500,00 €	168 750,00 €	253 125,00 €	379 687,50 €
Services (National market)	30 000,00 €	150 000,00 €	206 250,00 €	283 593,75 €	389 941,41 €	536 169,43 €
Services (Internationl market)	0,00 €	150 000,00 €	234 375,00 €	366 210,94 €	572 204,59 €	894 069,67 €
Other incomes	2 750,00 €	3 437,50 €	4 296,88 €	5 371,09 €	6 713,87 €	8 392,33 €
Total of incomes	47 750,00 €	415 937,50 €	608 984,38 €	894 824,22 €	1 319 470,21 €	1 952 361,30 €
Costs						
Investments	23 150,00 €	0,00 €	0,00 €	0,00 €	0,00 €	0,00 €
Loan Payments	0,00 €	0,00 €	0,00 €	0,00 €	0,00 €	0,00 €
Buildings + Renting	0,00 €	1 200,00 €	18 000,00 €	18 000,00 €	18 000,00 €	18 000,00 €
Salaries	56 700,00 €	126 000,00 €	216 000,00 €	270 000,00 €	360 000,00 €	360 000,00 €
Equipments	1 000,00 €	2 000,00 €	3 000,00 €	2 000,00 €	2 000,00 €	0,00 €
Advertising	12 000,00 €	36 000,00 €	12 000,00 €	12 000,00 €	12 000,00 €	12 000,00 €
Supplies	4 000,00 €	24 000,00 €	65 625,00 €	95 859,38 €	140 244,14 €	205 491,94 €
Travel	22 000,00 €	27 000,00 €	17 000,00 €	17 000,00 €	17 000,00 €	17 000,00 €
Vehicles	0,00 €	0,00 €	0,00 €	0,00 €	0,00 €	0,00 €
Telephone and Communications	2 400,00 €	3 600,00 €	3 780,00 €	3 969,00 €	4 167,45 €	4 375,82 €
Miscellaneous services	2 000,00 €	2 100,00 €	5 000,00 €	5 250,00 €	5 512,50 €	5 788,13 €
Total of costs	123 250,00 €	221 900,00 €	340 405,00 €	424 078,38 €	558 924,09 €	622 655,89 €
Net Income	-75 500,00 €	194 037,50 €	268 579,38 €	470 745,84 €	760 546,12 €	1 329 705,41 €
Discounted net income	-75 500,00 €	168 728,26 €	203 084,59 €	309 523,03 €	434 844,71 €	661 098,59 €

Figure 4.7   Projected cash flows for optimistic scenario.

Table 4.5   Financial indicators for the optimistic scenario

NPV	1,701,779.20 €
IRR	304.72%
Payback	1.39 (1 year and 5 months)

Table 4.6   Ratios for the standard scenario

Year	2017	2018	2019	2020	2021	2022
GPM	60%	60%	60%	60%	60%	60%
ROE	−326.13%	603.49%	1160.17%	2033.46%	3285.30%	5743.87%
ROI	−61.26%	87.44%	78.90%	111.00%	136.07%	213.55%

The GPM, like in the pessimistic scenario, remains at 60% due to the same reasons. The ROE increases significantly along the years, and the same situation happens also with ROI, with only a slight reduction from 2018 to 2019.

4.2.3 Sensibility Analysis

In order to perform a sensitivity analysis, we will look in more detail for the critical variables in our business. In terms of incomes, the majority of revenues come from product sales and services. Looking into costs, we realize that salaries are our most significant cost.

<u>Product sales and services</u>

We changed the estimated incomes from sales and services (domestic and international markets) from −25% to 25%. The impact of this change in terms of NPV, IRR, and payback is given in Tables 4.7 and 4.8.

It is possible to realize that NPV is always negative when sales and services are reduced 10% or more. In such situation, it is not possible to calculate the IRR and payback, because it is not possible to recover the initial investment. However, the NPV remains positive even when we decrease the

Table 4.7   Sensibility analysis by decreasing sales & services

Sales & Service	−25%	−20%	−15%	−10%	−5%
NPV	−495 703,29 €	−384 313,51 €	−255 873,18 €	−108 055,22 €	61 689,21 €
IRR	Not applied	Not applied	Not applied	Not applied	65,72%
Payback	Not applied	Not applied	Not applied	Not applied	1,82

Table 4.8   Sensibility analysis by increasing sales & services

Sales & Service	5%	10%	15%	20%	25%
NPV	478 325,19 €	731 518,69 €	1019 263,81 €	1345 377,39 €	1713 961,77 €
IRR	166,57%	205,64%	243,18%	280,27%	317,47%
Payback	1,59	1,52	1,46	1,41	1,37

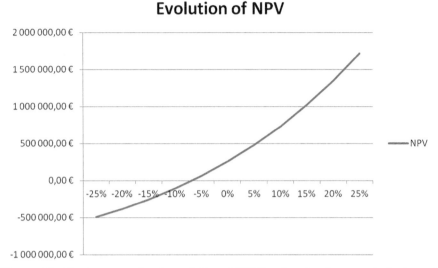

Figure 4.8   Graph representing the evolution of NPV by changing the amount of sales and services.

amount of sales and services in 5%. On the other hand, when we increase the number of sales and services, the NPV and IRR become more attractive. However, the payback does not suffer significant changes, because the cash flow does not change too much in the first year of the company's activity.

Finally, Figure 4.8 presents a graph of the evolution of NPV when we change the amount of sales and services.

Salaries

We changed also the amount paid in salaries. We considered a reduction up to −25% and an increment of +25%. The impact of this change in terms of NPV, IRR, and payback is given in Tables 4.9 and 4.10.

It is possible to realize that the NPV is still always possible, even when we increase the amount spent on salaries by 25%. The IRR adopts a similar behavior when compared to NPV. When looking for payback, we realize that it changes significantly from a period of 1.45 years to 2.11 years. The changes

Table 4.9   Sensibility analysis by decreasing the amount spent on salaries

Salaries	−25%	−20%	−15%	−10%	−5%
NPV	479 125,39 €	434 528,57 €	389 931,75 €	345 334,93 €	300 738,11 €
IRR	217,19%	196,54%	177,06%	158,53%	140,74%
Payback	1,45	1,49	1,53	1,58	1,63

Table 4.10   Sensibility analysis by increasing the amount spent on salaries

Salaries	5%	10%	15%	20%	25%
NPV	211 544,47 €	166 947,65 €	122 350,82 €	77 754,00 €	33 157,18 €
IRR	106,58%	89,68%	72,43%	54,16%	33,61%
Payback	1,76	1,83	1,91	2	2,11

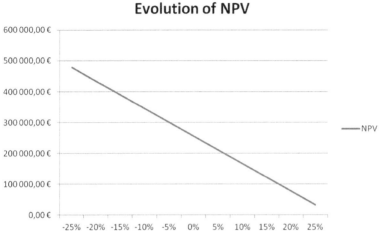

Figure 4.9   Graph representing the evolution of NPV by changing the amount spent in salaries.

on payback are more significant when we increase the amount paid on salaries rather than we reduced it.

Finally, Figure 4.9 presents a graph of the evolution of NPV when we change the amount spent on salaries.

4.3 Scenario II – AuditExpert

4.3.1 Key Objectives and Financial Review

Our financial goal is to maximize the NPV and IRR associated with the business. We intend to have an NPV higher than 100,000 in the first 6 years of operation and an IRR higher than 50% in the same period of time.

The start-up costs are determined in Figure 4.10. The initial investment in the project corresponds to around 100,000 €, composed of the following structure of costs:

- Leasehold improvements – a monthly fee of 200 € paid to the science park during the first year of activity;

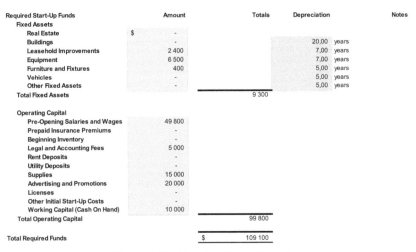

Required Start-Up Funds	Amount	Totals	Depreciation		Notes
Fixed Assets					
Real Estate	$ -				
Buildings			20,00	years	
Leasehold Improvements	2 400		7,00	years	
Equipment	6 500		7,00	years	
Furniture and Fixtures	400		5,00	years	
Vehicles	-		5,00	years	
Other Fixed Assets	-		5,00	years	
Total Fixed Assets		9 300			
Operating Capital					
Pre-Opening Salaries and Wages	49 800				
Prepaid Insurance Premiums	-				
Beginning Inventory	-				
Legal and Accounting Fees	5 000				
Rent Deposits	-				
Utility Deposits	-				
Supplies	15 000				
Advertising and Promotions	20 000				
Licenses	-				
Other Initial Start-Up Costs	-				
Working Capital (Cash On Hand)	10 000				
Total Operating Capital		99 800			
Total Required Funds	$	109 100			

Figure 4.10   Required financial capital.

- Equipment – cost of three laptops, router, and external storage devices;
- Furniture – office supplies that could be needed;
- Pre-opening salaries and wages – we considered a gross salary of 1500 € for three persons during 3 months. A gross salary of 1500 €, attending to Portuguese final taxes, will involve a total cost of 2500 € by month for each employee. We also estimate a total cost of 800 € per month for two internships, which already include a 50% subsidy supported by IEFP (Portuguese employment agency);
- Supplies – for the initial period of 3 months, we estimate that we will need 60 hardware boxes with a total cost of 15,000 €, considering that each unit has a cost of 250 €;
- Advertising and promotions – we estimate an initial cost of 20,000 € in advertising, which includes participation in two international fairs, participation in job fairs, and presence in technical magazines;
- Working capital – an estimated value of 10,000 € as working capital.

Each of the three founders will invest 5000 € in the business. The remaining necessary budget will come from risk capital funds (50,000 €) and banking entity (44,100 €) as indicated in Figure 4.11.

The projected cash flows are given in Figure 4.12. The analysis period considered is between 2017 and 2022, where the cash flows are estimated during 6 years. In incomes section, we include earnings from product and services sales from national and international markets; in cost section, we

Sources of Funding	Amount	Totals	Loan Rate	Term in Months	Monthly Payments
Owner's Cash Injection	13,75%	15 000			
Outside Investors	45,83%	50 000			
Additional Loans or Debt					
Commercial Loan	40,42%	44 100	9,00%	84,00	$709,53
Commercial Mortgage	0,00%	-	9,00%	240,00	$0,00
Total Sources of Funding	100,00%	$ 109 100			$709,53

Figure 4.11   Sources of funding.

	2017	2018	2019	2020	2021	2022
Incomes						
Product Sales (National market)	130 000,00€	136 500,00€	143 325,00€	150 491,25€	158 015,81€	165 916,60€
Product Sales (International market)	0,00€	0,00€	212 500,00€	265 625,00€	332 031,25€	415 039,06€
Services (National market)	0,00€	12 000,00€	12 600,00€	13 860,00€	15 246,00€	16 770,60€
Services (Internationl market)	0,00€	0,00€	30 000,00€	37 500,00€	46 875,00€	58 593,75€
Other incomes	0,00€	6 000,00€	9 000,00€	13 500,00€	20 250,00€	30 375,00€
Total of incomes	130 000,00€	154 500,00€	407 425,00€	480 976,25€	572 418,06€	686 695,02€
Costs						
Investments	109 100,00€	0,00€	0,00€	0,00€	0,00€	0,00€
Loan Payments	8 514,36€	8 514,36€	8 514,36€	8 514,36€	8 514,36€	8 514,36€
Risk capital payments	0,00€	0,00€	20 000,00€	20 000,00€	60 000,00€	0,00€
Buildings +Renting	0,00€	2 400,00€	18 000,00€	18 900,00€	19 845,00€	20 837,25€
Salaries	49 800,00€	109 200,00€	150 000,00€	157 500,00€	165 375,00€	173 643,75€
Equipments	0,00€	2 000,00€	6 000,00€	2 000,00€	2 100,00€	2 205,00€
Advertising	0,00€	20 000,00€	30 000,00€	30 000,00€	22 500,00€	22 500,00€
Supplies	0,00€	15 750,00€	88 956,25€	104 029,06€	122 511,77€	145 238,92€
Travel	18 000,00€	19 800,00€	21 780,00€	23 958,00€	26 353,80€	28 989,18€
Vehicles	0,00€	0,00€	43 600,00€	3 780,00€	3 969,00€	4 167,45€
Telephone and Communications	3 000,00€	3 150,00€	4 725,00€	4 961,25€	5 209,31€	5 469,78€
Miscellaneous services	2 400,00€	2 520,00€	2 646,00€	2 778,30€	2 917,22€	3 063,08€
Total of costs	190 814,36€	183 334,36€	394 221,61€	376 420,97€	439 295,45€	414 628,76€
Net Income	-60 814,36€	-28 834,36€	13 203,39€	104 555,28€	133 122,61€	272 066,26€
Discounted net income	-60 814,36€	-25 073,36€	9 983,66€	68 746,79€	76 113,28€	135 265,01€

Figure 4.12   Projected cash flows for standard scenario.

include rubrics such as investments, monthly loan payments, advertising, and supplies. In more detail, we can look to the income elements:

- Product sales from national market – in 2017, we predict sales of 30 (max: 10 users), 20 (max: 50 users), and 10 (max: 250 users) of IT security unified platforms. We do not predict to sell any IT security unified security platforms in the segment of no limit number of users. We estimate an increase of 5% in revenue per year for the following years;

- Product sales from international market – revenues from international markets will only appear in 2019. In this year, we estimate sales of 50 (max: 10 users), 25 (max: 50 users), 10 (max: 250 users), and 5 (no limit number of users) in each segment. We estimate an increase of 20% in revenue per year for the following years;

- Services' revenues from national market – revenues from services will only appear in 2018. For that year, we estimate 12 sales of IT audit services, which give approximately one audit service per month. We estimate an increase of 5% in revenue per year for the following years;

- Services from international market – like product sales in the international market, revenues from international markets will only appear in 2019. In this year, we estimate 30 sales of IT audit services, which give approximately less than 3 audit services per month. We estimate an increase of 25% in revenue per year for the following years;
- Other incomes – in this section, we include additional remote and local technical support. We estimate to have a revenue of 500 € per month for 2019. This revenue will potentially increase by 50% for the following years, considering the number of IT security unified platforms installed on clients.

Our outcomes are organized as follows:

- Investments – we only consider an initial investment of 109,100 €. Other investments needed along the time will be placed in other sections, such as advertising, equipment, buildings, or vehicles;
- Loan payments – we estimate a monthly fee of 709.53 € during 84 months (7 years);
- Risk capital payments – a total fee of 100,000 € must be paid during the period of business to risk capital company. The first two payments in the value of 20,000 € must be paid in 2019 and 2020. The remaining fee (60,000 €) must be paid in 2021;
- Buildings + Renting – for 2018, we estimate a similar cost compared to the previous year in terms of renting a space in the science park. In 2019, we intend to have our own installations and this will cost us an amount of 1500 € per month. After 2019, we estimate that our costs in terms of buildings and renting will increase 5% per year;
- Salaries – for 2017, we estimate to have 3 employees with contract and two additional internships. In 2018, we will intend to have all employees with a contract with a total of five employees. In 2019, our structure of costs in terms of salaries will change a little, by the recruitment of three additional employees. Then, in 2019, we estimate to have two classes of employees: one with an average salary of 2500 € per month (3 employees); another with an average salary of 1000 € per month (5 employees). After 2019, we predict to have an increase of 5% in salaries;
- Equipment – a fee of 2000 €in 2018 is estimated as enough to support the costs of equipment for our collaborators. This fee will increase 3 times, with the recruitment of new employees in 2019. For 2020, we

estimate to have a basic fee of 2000 €, with an increase of 5% for the following years;

- Advertising – our initial investment includes all costs in advertising for 2017. The same amount of expenses is predicted for 2018. Then, with our presence in the international market, we expect to have an increase of 50% in our costs in 2019 and 2020. For 2021 and 2022, we expect to decrease our costs by 25% compared to previous years;
- Supplies – the cost of suppliers represents approximately 25% of the costs of our sales;
- Travel – we estimate a cost of 1500 € per month in traveling to our clients. We also estimate an increase of 10% per year in these costs;
- Vehicles – during the first two years of activity, we do not plan to have an investment in vehicles. This situation will change in 2019, where we intend to buy two commercial cars with a cost of 20,000 €, each one. Additionally, we reserve a fee of 300 € per month for maintenance costs. For 2020, we only include the maintenance costs with an increase of 5% per year. The same procedure applies for the following years;
- Telephone and communications – for 2017, we estimate to have a cost of 250 € per month. These costs will increase by 5% in 2018 and by 50% for 2019, due to the presence in international markets. For the next years, we plan to have an increase of 5%;
- Miscellaneous services – other type of costs associated with services has an estimate of 200 € per month, which represents 2400 € per year. This amount will increase to 5% for next years.

4.3.2 Viability Analysis

The viability analysis of a given business can be assessed by considering three scenarios: standard, pessimist, and optimist. The premises of each scenario will be presented and we calculate the NPV, IRR, and payback. We considered a discount rate of 15% in all scenarios.

Standard Scenario

The standard scenario uses the same cash flow projections available in Figure 4.12. The values of NPV, IRR, and payback are available in Table 4.11. Moreover, the analysis of the most relevant ratios is given in Table 4.12.

Our analysis indicates an NPV higher than 200,000 € for 6 years and an IRR of 58.94%. NPV higher than 0 and an IRR higher than the discount rate of 15% indicates that the project is profitable. Besides that the payback is estimated at 3 years and 9 months.

Table 4.11   Financial indicators for the standard scenario

NPV	204,221.03 €
IRR	58.94%
Payback	3.79 (3 years and 9 months)

Table 4.12   Ratios for the standard scenario

Year	2017	2018	2019	2020	2021	2022
GPM	25%	25%	25%	25%	25%	25%
TDA	78.15%	73.98%	69.81%	55.85%	41.89%	8.34%
ROE	−55.74%	−26.43%	8.71%	69.00%	87.85%	179.55%
ROI	−31.87%	−15.73%	3.35%	27.78%	30.30%	65.62%

The gross profit margin (GPM) is estimated as being 25% in all considered period. The total debt to total assets (TDA) will be decreasing along the period of time, because part of the debt will be paid to the bank and the risk capital entity. It is important to highlight that the total term debt includes the amount of the bank loan plus the total of interest plus risk capital debt. In total, this value is equivalent to 159,600 €. The ROE and ROI become positive only in 2019.

Pessimistic Scenario

The projected cash flows are given in Figure 4.13. For the construction of the pessimistic scenario, we introduced the following changes:

- The products sales and services, both in national and international markets, were reduced by 10% for all considered years;
- The monthly loan payments increase at 20%;
- Costs in advertising increase 10% and the costs in suppliers increase 20% in all considered period.

The pessimistic scenario will have a significant impact on our business. The yearly cash flow will become only positive for 2020. The values of NPV, IRR, and payback are available in Table 4.13. The analysis of the most relevant ratios is given in Table 4.14.

The NPV will remain positive even considering the pessimistic scenario. IRR value is slightly higher than our established discounted rate. The payback will be increased significantly for 5 years and 10 months.

The GPM remains unalterable compared to the standard scenario. The TDA is very high until 2021, because the total amount of debt is very high when compared to the NPV offered by the business. The ROE and ROI are negative in the first 3 years of operation.

	2017	2018	2019	2020	2021	2022
Incomes						
Product Sales (National market)	117 000,00€	122 850,00€	128 992,50€	135 442,13€	142 214,23€	149 324,94€
Product Sales (International market)	0,00€	0,00€	191 250,00€	239 062,50€	298 828,13€	373 535,16€
Services (National market)	0,00€	10 800,00€	11 340,00€	12 474,00€	13 721,40€	15 093,54€
Services (Internationl market)	0,00€	0,00€	27 000,00€	33 750,00€	42 187,50€	52 734,38€
Other incomes	0,00€	6 000,00€	9 000,00€	13 500,00€	20 250,00€	30 375,00€
Total of incomes	117 000,00€	139 650,00€	367 582,50€	434 228,63€	517 201,26€	621 063,01€
Costs						
Investments	109 100,00€	0,00€	0,00€	0,00€	0,00€	0,00€
Loan Payments	10 217,23€	10 217,23€	10 217,23€	10 217,23€	10 217,23€	10 217,23€
Risk capital payments	0,00€	0,00€	20 000,00€	20 000,00€	60 000,00€	0,00€
Buildings +Renting	0,00€	2 400,00€	18 000,00€	18 900,00€	19 845,00€	20 837,25€
Salaries	49 800,00€	109 200,00€	150 000,00€	157 500,00€	165 375,00€	173 643,75€
Equipments	0,00€	2 000,00€	6 000,00€	2 000,00€	2 100,00€	2 205,00€
Advertising	0,00€	22 000,00€	33 000,00€	33 000,00€	24 750,00€	24 750,00€
Supplies	0,00€	18 900,00€	96 072,75€	112 351,39€	132 312,71€	156 858,03€
Travel	18 000,00€	19 800,00€	21 780,00€	23 958,00€	26 353,80€	28 989,18€
Vehicles	0,00€	0,00€	43 600,00€	3 780,00€	3 969,00€	4 167,45€
Telephone and Communications	3 000,00€	3 150,00€	4 725,00€	4 961,25€	5 209,31€	5 469,78€
Miscellaneous services	2 400,00€	2 520,00€	2 646,00€	2 778,30€	2 917,22€	3 063,08€
Total of costs	192 517,23€	190 187,23€	406 040,98€	389 446,17€	453 049,27€	430 200,75€
Net Income	-75 517,23€	-50 537,23€	-38 458,48€	44 782,46€	64 151,99€	190 862,27€
Discounted net income	-75 517,23€	-43 945,42€	-29 080,14€	29 445,19€	36 679,11€	94 892,28€

Figure 4.13   Projected cash flows for pessimistic scenario.

Table 4.13   Financial indicators for the pessimistic scenario

NPV	12,473.79 €
IRR	17.52%
Payback	5.87 (5 years and 10 months)

Table 4.14   Ratios for the pessimistic scenario

Year	2017	2018	2019	2020	2021	2022
GPM	25%	25%	25%	25%	25%	25%
TDA	1279.48%	1197.57%	1115.66%	873.42%	631.17%	68.25%
ROE	−69.22%	−46.32%	−35.25%	29.55%	42.34%	125.96%
ROI	−39.23%	−26.57%	−9.47%	11.50%	14.16%	44.37%

Optimistic Scenario

The projected cash flows are given in Figure 4.14. The optimistic scenario considers the following premises:

- The products sales and services increased to 10% for all considered years;
- Costs in advertising also increased 10%, considering that the increment in the number of sales also originates an increase in advertising in the same proportion;

	2017	2018	2019	2020	2021	2022
Incomes						
Product Sales (National market)	143 000,00 €	150 150,00 €	157 657,50 €	165 540,38 €	173 817,39 €	182 508,26 €
Product Sales (International market)	0,00 €	0,00 €	233 750,00 €	292 187,50 €	365 234,38 €	456 542,97 €
Services (National market)	0,00 €	13 200,00 €	13 860,00 €	15 246,00 €	16 770,60 €	18 447,66 €
Services (Internationl market)	0,00 €	0,00 €	33 000,00 €	41 250,00 €	51 562,50 €	64 453,13 €
Other incomes	0,00 €	6 000,00 €	9 000,00 €	13 500,00 €	20 250,00 €	30 375,00 €
Total of incomes	143 000,00 €	169 350,00 €	447 267,50 €	527 723,88 €	627 634,87 €	752 327,02 €
Costs						
Investments	109 100,00 €	0,00 €	0,00 €	0,00 €	0,00 €	0,00 €
Loan Payments	8 514,36 €	8 514,36 €	8 514,36 €	8 514,36 €	8 514,36 €	8 514,36 €
Risk capital payments	0,00 €	0,00 €	20 000,00 €	20 000,00 €	60 000,00 €	0,00 €
Buildings +Renting	0,00 €	2 400,00 €	18 000,00 €	18 900,00 €	19 845,00 €	20 837,25 €
Salaries	49 800,00 €	109 200,00 €	150 000,00 €	157 500,00 €	165 375,00 €	173 643,75 €
Equipments	0,00 €	2 000,00 €	6 000,00 €	2 000,00 €	2 100,00 €	2 205,00 €
Advertising	0,00 €	22 000,00 €	33 000,00 €	33 000,00 €	24 750,00 €	24 750,00 €
Supplies	0,00 €	12 600,00 €	78 281,50 €	91 545,58 €	107 810,35 €	127 810,25 €
Travel	18 000,00 €	19 800,00 €	21 780,00 €	23 958,00 €	26 353,80 €	28 989,18 €
Vehicles	0,00 €	0,00 €	43 600,00 €	3 780,00 €	3 969,00 €	4 167,45 €
Telephone and Communications	3 000,00 €	3 150,00 €	4 725,00 €	4 961,25 €	5 209,31 €	5 469,78 €
Miscellaneous services	2 400,00 €	2 520,00 €	2 646,00 €	2 778,30 €	2 917,22 €	3 063,08 €
Total of costs	190 814,36 €	182 184,36 €	386 546,86 €	366 937,49 €	426 844,04 €	399 450,09 €
Net Income	-47 814,36 €	-12 834,36 €	60 720,64 €	160 786,39 €	200 790,83 €	352 876,93 €
Discounted net income	-47 814,36 €	-11 160,31 €	45 913,53 €	105 719,66 €	114 802,81 €	175 442,20 €

Figure 4.14 Projected cash flows for optimistic scenario.

- Costs of suppliers are reduced by 20%, considering that we got a significant discount from our supplier.

The optimistic scenario guarantees positive cash flows in the third year of our activity. Particularly important is the significant income that we get for 2021 and 2022. However, it is also relevant to highlight that projections in the fifth and sixth years of activity have always been a high-level uncertainty. The values of NPV, IRR, and payback are available in Table 4.15. Furthermore, the analysis of the most relevant ratios is given in Table 4.16.

Table 4.15 Financial indicators for the optimistic scenario

NPV	382,903.52 €
IRR	105.18 %
Payback	3.20 (3 years and 2 months)

Table 4.16 Ratios for the optimistic scenario

Year	2017	2018	2019	2020	2021	2022
GPM	25%	25%	25%	25%	25%	25%
TDA	41.68%	39.46%	37.23%	29.79%	22.34%	4.45%
ROE	−43.83%	−11.76%	40.07%	106.11%	132.51%	232.88%
ROI	−25.06%	−7.04%	15.71%	43.82%	47.04%	88.34%

Both NPV and IRR will be increased compared to the standard scenario. The payback is reduced in around 7 months.

The GPM like in the previous scenario assumes always the same value. The TDA is close to half the value of the standard scenario. The ROE and ROI are both negative in the first 2 years of operation.

4.3.3 Sensibility Analysis

In order to perform a sensitivity analysis, we will look in more detail for the critical variables in our business. In terms of incomes, the majority of revenues come from product sales and services. Other incomes offer also a relevant source of income. On the other hand, the most significant costs are monthly loan payments, salaries, advertising, and cost of supplies.

Product sales and services

We changed the value of our estimated product sales and services in a scale between −25% and 25%. The impact of this change in terms of NPV, IRR, and payback is given in Tables 4.17 and 4.18.

The NPV becomes negative when the value of sales and services are reduced to −20% and −25%. The IRR has similar behavior compared to NPV and is below the discount rate of 15% for the same scenarios. It is also not possible to recover the performed initial investment for the scenario of −25%.

Finally, Figure 4.15 presents a graph of the evolution of NPV when we change the amount of sales and services.

Table 4.17 Sensibility analysis by decreasing sales & services

Sales & Service	−25%	−20%	−15%	−10%	−5%
NPV	−103 744,15 €	−42 151,11 €	19 441,93 €	81 034,96 €	142 628,00 €
IRR	Not applied	6,86%	18,81%	31,21%	44,45%
Payback	Not applied	5,44	5,23	4,74	4,17

Table 4.18 Sensibility analysis by increasing sales & services

Sales & Service	5%	10%	15%	20%	25%
NPV	265 814,07 €	327 407,10 €	389 000,14 €	450 593,17 €	512 186,21 €
IRR	75,25%	94,21%	117,08%	145,97%	184,85%
Payback	3,51	3,3	3,14	3	2,89

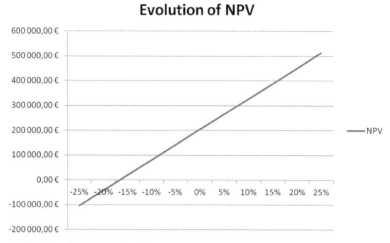

Figure 4.15   Graph representing the evolution of NPV by changing the amount of sales and services.

Other incomes

In this scenario, we changed the value of other incomes and we measure its impact in terms of NPV, IRR, and payback. Tables 4.19 and 4.20 summarize the effects of this change.

The NPV remains always positive for all considered changes in the other incomes rubric. In fact, the changes in the NPV are not very significant. A similar conclusion can be extracted when we look into IRR and payback.

Finally, Figure 4.16 presents a graph of the evolution of NPV when we change the amount of sales and services.

Table 4.19   Sensibility analysis by decreasing other incomes

Other Incomes	−25%	−20%	−15%	−10%	−5%
NPV	175 374,98 €	179 407,07 €	184 189,51 €	189 834,77 €	196 466,69 €
IRR	53,91%	54,67%	55,55%	56,54%	57,67%
Payback	3,88	3,86	3,84	3,82	3,81

Table 4.20   Sensibility analysis by increasing other incomes

Other Incomes	5%	10%	15%	20%	25%
NPV	213 246,06 €	223 703,11 €	235 767,13 €	249 627,28 €	265 487,47 €
IRR	60,36%	61,94%	63,69%	65,61%	67,71%
Payback	3,77	3,75	3,74	3,72	3,71

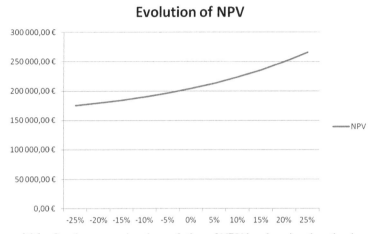

Figure 4.16 Graph representing the evolution of NPV by changing the other incomes.

Loan payments

We also wanted to analyze the consequences of the existence of changes in the interest rates. We considered a scenario that includes a maximum reduction and an increment in the monthly loan payment. Tables 4.21 and 4.22 summarize this scenario.

The NPV does not change significantly by the evolution of monthly loan payments. A similar conclusion can be drawn with the IRR and payback period.

Finally, Figure 4.17 presents a graph of the evolution of NPV when we change the monthly loan payments.

Table 4.21 Sensibility analysis by decreasing monthly loan payments

Monthly Loan Payments	−25%	−20%	−15%	−10%	−5%
NPV	213 484,99 €	211 632,20 €	209 779,40 €	207 926,61 €	206 073,82 €
IRR	61,99%	61,36%	60,75%	60,14%	59,53%
Payback	3,73	3,74	3,75	3,76	3,78

Table 4.22 Sensibility analysis by increasing monthly loan payments

Monthly Loan Payments	5%	10%	15%	20%	25%
NPV	203 219,68 €	202 216,32 €	201 216,97 €	200 215,61 €	199 214,26 €
IRR	58,81%	58,69%	58,56%	58,43%	58,31%
Payback	3,79	3,8	3,81	3,81	3,82

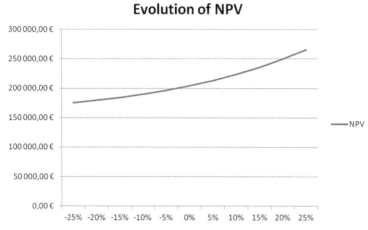

Figure 4.17 Graph representing the evolution of NPV by changing the monthly loan payments.

Salaries

The cost of salaries is our main expense. Therefore, we changed our estimated costs for this rubric and we measured its impact on the business profitability. Tables 4.23 and 4.24 summarize the obtained results.

All financial indicators (NPV, IRR, and payback) suffer a very significant impact. The NPV turns negative when we increase the cost of salaries in 20% or 25%. The IRR has a similar behavior compared to NPV. On the other hand, the payback increases significantly with the increase in the amount spent in salaries. In a scenario, when we increase the cost of salaries 20% or 25%, it is not possible to recover the initial investment.

Table 4.23 Sensibility analysis by decreasing the total cost of salaries

Salaries	−25%	−20%	−15%	−10%	−5%
NPV	427 752,15 €	390 250,21 €	349 437,20 €	305 028,51 €	256 726,62 €
IRR	113,09%	101,80%	90,97%	80,38%	69,79%
Payback	3,24	3,31	3,4	3,5	3,62

Table 4.24 Sensibility analysis by increasing the total cost of salaries

Salaries	5%	10%	15%	20%	25%
NPV	147 188,32 €	85 292,09 €	18 183,02€	−54 501,19 €	−133 135,76 €
IRR	47,43%	34,68%	19,55%	Not applied	Not applied
Payback	4,03	4,46	5,47	Not applied	Not applied

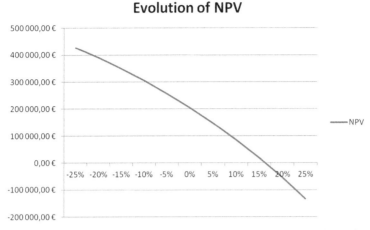

Figure 4.18   Graph representing the evolution of NPV by changing the total costs in salaries.

Finally, Figure 4.18 presents a graph of the evolution of NPV when we increase the total costs in salaries.

Advertising

The costs in advertising do not have a major impact on our business profitability, compared to the costs of employees. Tables 4.25 and 4.26 summarize the obtained results.

The impact on all financial indicators is not very high. When we vary the costs in advertising from −25% to +25%, the NPV decreases in around 90,000 € and the IRR also decreases around 20%. Similar behavior happens to payback that increases from 3.59 to 4.10 years.

Table 4.25   Sensibility analysis by decreasing the costs in advertising

Advertising	−25%	−20%	−15%	−10%	−5%
NPV	241 027,58 €	234 703,67 €	227 879,10 €	220 535,82 €	212 655,82 €
IRR	67,22%	65,74%	64,18%	62,52%	60,78%
Payback	3,59	3,63	3,66	3,7	3,74

Table 4.26   Sensibility analysis by increasing the costs in advertising

Advertising	5%	10%	15%	20%	25%
NPV	195 234,44 €	185 614,99 €	175 407,65€	164 573,39 €	153 094,16 €
IRR	57,00%	54,95%	52,79%	50,52	48,12
Payback	3,84	3,39	3,95	4,02	4,1

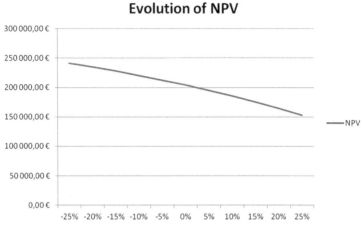

Figure 4.19   Graph representing the evolution of NPV by changing the costs in advertising.

Finally, Figure 4.19 presents a graph of the evolution of NPV when we increase the costs in advertising.

Suppliers

The costs of supplies are our second main expense. However, the impact in the profitability of our business is not so high like it happened for the cost of employees. Tables 4.27 and 4.28 summarize the obtained results.

Table 4.27   Sensibility analysis by decreasing the cost of supplies

Cost of Supplies	−25%	−20%	−15%	−10%	−5%
NPV	277 125,05 €	262 544,24 €	247 963,44 €	233 382,64 €	218 801,84 €
IRR	73,02%	70,26%	67,47%	64,56%	61,81%
Payback	3,51	3,56	3,61	3,66	3,72

Table 4.28   Sensibility analysis by increasing the cost of supplies

Cost of Supplies	5%	10%	15%	20%	25%
NPV	189 640,23 €	175 059,43 €	160 478,62€	145 897,82 €	131 317,02 €
IRR	56,04%	53,11%	50,15%	47,15	44,12
Payback	3,86	3,94	4,03	4,13	4,23

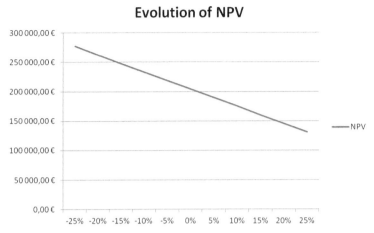

Figure 4.20 Graph representing the evolution of NPV by changing the cost of supplies.

The impact on our financial indicators is moderate. The NPV decreases around 140,000 €, and the IRR decreases around 30%. The payback increases in almost 1 year, from 3.51 years to 4.23 years. This situation can be explained by the fact that the cost of supplies is more relevant in the last 2 years of the business activity.

Finally, Figure 4.20 presents a graph of the evolution of NPV when we increase the cost of supplies.

5

Prototype Description

Overview

The purpose of this chapter is to provide an overview of the process of building the IT application. This information is relevant to both promoters and potential investors in the business, since with this chapter, we intend to emphasize the degree of innovation offered by the application. For this purpose, we first describe the functional and non-functional requirements of the prototype, present the physical and logical architecture of the application, and perform a modeling of the most relevant functionalities offered by the prototype. Finally, we use two scenarios (*TourMCard* and *AuditExpert*) to demonstrate their use, considering different types of functionalities and technological architectures.

5.1 Theoretical Foundations

5.1.1 Prototype Features

A first initial step is to present the prototype features of the IT business proposal. For that, we need to present all the functionalities offered by the product and indicate what will be implemented in the prototype. Of course, the prototype does not need to implement all the functionalities of the product, but only part of them could give a good overview about the characteristics and distinctive features offered by our product.

In order to present the product features, we shall use the unified modeling language (UML) in its version 2.0. A modeling standard language, such as UML, can be used to design software applications. Using a model, those responsible for a software development project's success can assure themselves that business functionality is complete and correct, end-user needs are met, and program design supports requirements for scalability, robustness, security, extendibility, and other characteristics, before implementation in code renders changes that are difficult and expensive to make.

UML offers several diagrams for different purposes. One of the most basic UML diagrams is the **use case diagram**. This diagram presents what the system does from the user's point of view. Thus, it describes the main features of the system and the interaction of these features with the users of the same system. However, the use case diagrams do not give any technical detail how the system is implemented.

The use case diagrams are generically composed of three parts:

- Actors: represent external entities that interact with the system during its execution. Basically, the interaction of actors in the system occurs through the exchange of messages. The external entities represented by the actors can be people and their roles in system, devices, or software applications;
- Use cases: represent a task or functionality performed by the actor. It must include a verb in order to express the notion of action;
- Communications: what binds an actor to a use case. The relationships in a use case diagram can involve two actors, two use cases, or an actor and a use case.

In order to express relationships between two use cases, we can use the notation "include" and "extend". "Include" means that the execution of the parent use case involves mandatorily the execution of the child use case. On the other hand, "extend" means that the child use case will only be executed in some situations. Therefore, the execution of the child use case is optional.

The use case diagram can also express a generalization relationship between two actors or two use cases. A generation relationship is a structural relationship between a more general use case and a more specific one. The most general use case represents the generic case, whose service applies to several situations. The most specific use case represents the application of the most general use case in a particular situation, including additional elements or extending this case. Seen in another perspective, the most general use case is a generalization or abstraction of a more specific use case. A similar reasoning can be applied to the concept of generalization between actors.

A software project typically contains a single use case diagram, which describes the set of features offered by the system. For larger or more complex systems, however, it is possible to construct several use case diagrams drawn from the decomposition of the main diagram. The decomposition of a use case diagram can be done in UML using the concept of a package.

A package can be seen as an encapsulator that is used to separate or group elements of the project. A good approach is to create a first use case diagram containing all the packages, and then take each packet and expand it into a new diagram. Therefore, it is possible to build a hierarchy with various levels of decomposition depending on the size of the system and the number of use cases and actors.

UML provides a full set of diagrams that can be used to better understand a complex system. These diagrams are generally made in an incremental and iterative way and can be divided into two categories: (i) structural diagrams and (ii) behavioral diagrams. Structural diagrams represent the static aspect of the system that typically corresponds to its main structure. On the other hand, behavioral diagrams basically capture the dynamic aspect of a system.

The most common UML structural diagrams are:

- Class diagrams – it is one of the most used diagrams, especially when modeling the structure of a relational database. It allows the visualization of the classes used by the system and how they are related. A class represents a description of a set of objects that share the same attributes, operations, relationships, and semantics;
- Deployment diagrams – it is also a very common diagram that is particularly used to represent the physical architecture of the system. Deployment diagrams are used for describing the hardware components where software components are deployed;
- Component diagrams – it is also a very useful diagram used to represent the logical architecture of the system. Component diagrams show the structure of the system, which describes the software components, their interfaces, and their dependencies.

On the other hand, the most common UML behavioral diagrams are:

- Sequence diagrams – represent the interactions between objects through messages that are exchanged between them, further specifying which is the appropriate temporal chaining (temporal sequence);
- Activity diagrams – provide a visualization of the behavior of a system for describing the sequence of actions in a process. Activity diagrams are similar to flowcharts because they show the flow between actions in an activity. In an activity diagram, we can represent parallel or simultaneous flows and alternative flows;
- State diagrams – present the different behaviors of a system, a component or an object, through all the states and transitions that can take over time. The use of state diagrams is not so common as sequence

or activity diagrams, because they are only useful when there are objects in a system that can have different states and transitions among them.

Also important to consider in the development of a software prototype is the non-functional requirements. These represent the characteristics and internal aspects of the system, specifically involving the technical part. Unlike functional requirements, these requirements are not explicitly exposed by the client, but must be implicitly understood by the business analyst and developer. Non-functional requirements define constraints in the product to be developed, in the process of development, or there are just external factors that the product must attend (Almeida et al., 2017). In any situation, it is important to highlight that non-functional requirements may have a high impact on software quality.

In fact, non-functional requirements are so important as functional requirements or business rules. Unfortunately, it is common to see that many software engineers and companies do not take this seriously and, therefore, many projects fail due to this situation. There are typically two reasons that justify the little importance that non-functional requirements receive in most projects: (i) users do not know what a non-functional requirement is and (ii) non-functional requirements are difficult to estimate.

Reifer and Boehm (2006) provide a good overview on the organization of non-functional requirements in three groups: (i) process requirements; (ii) product requirements; and (iii) external requirements. Figure 5.1 presents this classification.

The non-functional requirements can be divided into several areas. The most common and relevant non-functional requirements are:

- Accessibility – refers to the design of products, devices, services, or environments for people who experience disabilities;
- Availability – the system may be available to the end-user;
- Backup – existence of multiple copies of the system and/or data in order to prevent data loss and restore it when needed;
- Documentation – creation of user guides, quick reference guides, online help, or contextualized help that could guide the end-user through the use of the application;
- Interoperability – needs of system integration with other systems, integration with APIs, components, Web services, or external databases;
- Legal – there may be legal issues involving privacy of information, intellectual property rights, export of restricted technologies, etc.;
- Modifiability – requirements about the effort required to make changes in the software;

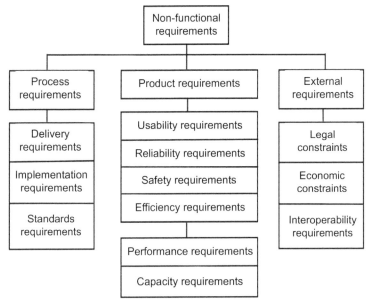

Figure 5.1 Classification of non-functional requirements.

- Performance – requirements about resources required, response time, transaction rates, throughput, benchmark specifications, or anything else having to do with performance;
- Portability – the effort required to move the software to a different target platform, namely to mobile devices;
- Reliability – requirements about how often the software fails and tolerance to non-expected outputs;
- Security – one or more requirements about protection of your system and its data;
- Standardization – use of open standards that could facilitate the software development or exchange of data;
- Usability – requirements about how difficult it will be to learn and operate the system.

In addition to the definition of the non-functional requirements, a set of metrics may be established to test and measure the degree of compliance with the non-functional requirements. Table 5.1 provides a suggestion of three metrics that can be used for each class of non-functional requirements. A good metric should try to be objective, quantitative, easy to obtain, and repeatable.

Table 5.1   Examples of metrics for non-functional requirements

Property	Metrics
Accessibility	1. Number of pictures without an "alt" attribute; 2. Web Accessibility Barrier (WAB); 3. Web Accessibility Quality Metric (WAQM).
Availability	1. Percentage of time that an application is available; 2. Percentage of calls abandoned while waiting to be answered; 3. Mean Time To Recover (MTTR)
Backup	1. Number of application backups per month; 2. Number of data backups per month; 3. Mean time to recover to a previous software version
Documentation	1. Existence and validation of software requirements report; 2. Existence and number of considered test cases; 3. Existence and number of coding rules.
Interoperability	1. Compatibility of the IT technological platform; 2. Compatibility of transferred data; 3. Compatibility of software languages.
Legal	1. Protection of personal data; 2. In accordance with national and European legal framework; 3. Authorization from external entities (if needed).
Modifiability	1. Time needed to include a new functional requirement; 2. Time needed to rewrite the code in a new programming language; 3. Time needed to migrate data to a new database management system.
Performance	1. Processed transactions per second; 2. Response time to user input; 3. Time need to load a Web page.
Portability	1. Number of target systems; 2. Quality of content adaptation to new devices; 3. Amount of time needed to migrate to a new server.
Reliability	1. Mean Time Between Failures (MTBF); 2. Annualized Failure Rate (AFR); 3. Workload Rate Limit (WRL).
Security	1. Percentage of systems with of systems with formal risk assessments; 2. Percentage of weak passwords; 3. Number of identified risks and their severity.
Standardization	1. Number of vendors lock-in; 2. Number of open standards adopted; 3. Number of obtained certifications.
Usability	1. Number of adopted usability principles; 2. Success rate to perform specific tasks; 3. Time needed to complete specific tasks.

5.1.2 Technology Choice and Architecture

In order to present the technology choice of a prototype, we shall use UML deployment diagrams. The purpose of deployment diagrams can be described as: (i) visualize the hardware topology of a system; (ii) describe the hardware components used to deploy software components; and (iii) describe runtime processing nodes. Therefore, in a common sense, we can assume that UML deployments are particularly suitable to model the physical architecture.

UML deployment diagrams can be used to model embedded systems, client/server system, or fully distributed systems. In order to produce a deployment diagram, we need to perform the following tasks:

- Decide the objectives to the diagram;
- Add nodes to the diagram;
- Add communication associations to the diagram;
- Add other elements to the diagram, such as components or active objects (if required);
- Add dependencies between components and objects (if required).

There are generally two big groups of IT architectures:

- Client–server architectures – clear distinction between clients and servers, in which many clients request services from a centralized server;
- Distributed object architectures – no distinction between clients and servers. Any object on the system may provide and use services from other objects.

The client/server architecture may be implemented using a layered application architecture approach. In a two-tier model, we can use thin or flat clients. In a thin model, all of the application processing and data management is carried out on the server. The client is simply responsible for presenting the results to the user interface. A major disadvantage is that it places a heavy processing load on both the server and the network. In a fat-client model, the server is only responsible for data management. On the other hand, the software on the client implements the application logic and interacts with the system user. A major disadvantage is that it can become more complex than the thin model, where new versions of the application have to be installed on all clients.

The three-tier model is composed of three layers:

- Presentation layer – concerned with presenting the results of a computation to system users and by collecting user inputs;
- Business logic layer or application processing layer – concerned with providing and implementing application-specific functionality, as specified in the functional requirements;
- Data management layer – concerned with managing the system databases.

Table 5.2   Applications of client/server (C/S) architecture

Architectures	Applications
Two-tier C/S architecture with thin clients	Legacy system applications where separating application processing and data management is impractical.
	Computationally intensive applications such as compilers with little or no data management.
	Data-intensive applications (browsing and querying) with little or no application processing.
Two-tier C/S architecture with fat clients	Applications where application processing is provided by off-the-shelf software (e.g., Microsoft Excel) on the client.
	Applications where computationally intensive processing of data (e.g., data visualization) is required.
	Applications with relatively stable end-user functionality used in an environment with well-established system management.
Three-tier or multi-tier C/S architecture	Large-scale applications with hundreds or thousands of clients.
	Applications where both the data and the application are volatile.
	Applications where data from multiple sources are integrated.

Table 5.2 presents the main situations where it is more suitable to use a two-tier or three-tier model approach.

A distributed architecture offers no distinction between clients and servers. The object communication is done through a middleware system that is responsible for handling the communication. This architecture is a very open-system architecture that allows new resources to be added to it as required. Therefore, the system is flexible and scalable. On the other hand, the main disadvantage is associated with the high complexity when implementing simple services.

The most common models of distributed architectures are:

- CORBA – the Common Object Request Broker Architecture (CORBA) is an open distributed object computing infrastructure, standardized by the Object Management Group (OMG). This architecture lets programmers to automate many common network programming tasks, such as object registration, location, error-handling, or operation dispatching;
- Peer-to-peer (P2P) – decentralized architecture where computations may be carried out by any node in the network. The overall system is designed to take advantage of the computational power and storage of a large number of networked computers;
- Service-oriented architectures (SOA) – it is also a decentralized architecture based on the principle of Web services. A Web service

is a standard approach to make a reusable component available and accessible across the Web. The main advantage is that the service provision is independent of the application using the services. Services can be accessed using SOAP protocols and are built using an XML-based standard (WSDL).

On the other hand, the logical architecture of an IT application shall be modeled using a component diagram, where we show how the components of a solution are organized and integrated. The logical architecture must adopt the principles of software architecture design that is commonly introduced in software engineering fields. These principles include:

- Abstraction – can be formulated as the psychological capacity that humans have to focus on a certain level of a problem, without taking into account irrelevant details of the lower level. The abstraction should be used as problem-solving techniques in various engineering areas, such as software field;
- Modularization – a technique used in several areas of engineering to build a product that is made up of components and modules, which can be assembled or integrated. This technique allows the intellectual effort to build a program to be reduced. It also facilitates the process of compiling and running a program;
- Encapsulation – for the abstraction to be implemented more rigorously, each component (module) of the software must encapsulate all internal implementation details and make its interface visible only. The component interface should tell you what it does, what it needs to interconnect with other components, and what it can offer to the other components. This principle, also called information hiding, suggests that the components are designed in such a way that their internal elements are inaccessible to other components. Access should only be made through the interface. This ensures the integrity of the components as it prevents their elements from being altered by other components;
- Reuse – in addition to facilitating the development process by reducing intellectual effort or facilitating compilation, components can be reused in different software. A function that has been built for specific software can be reused in another software. The specific functionality of each component will be used to determine the overall functionality of the software. Software with different global functionalities can be built with some common specific components. Typical reusable components (data types, functions, or classes) are stored in other non-functional

components called libraries. Components can be embedded during compilation or during execution. In the first case, the components are in either the compile or the static link libraries, and in the second case, they are in the dynamic link libraries;

- Generalization – the construction of specific components or modules that facilitate the process of software development must follow another important principle called "generalization". For a software component to have utility in several different programs, it should be as generic as possible. For this, it needs to be built with the purpose of offering general-purpose services.

Additionally, the logical architecture shall also present the UML class diagram or the entity/relationship (E/R) model, if a relational database is used.

5.1.3 Prototype Modeling

A prototype is a concrete representation of part of all of an interactive system. It is like an interactive mockup that can have any degree of fidelity and provide a great deal of insight into the user interaction at various levels. They do not only let the company to test the feasibility and usability of the user interface design, but also lead to unexpected discoveries and innovations that may or may not take our project beyond its initial scope. Three benefits are widely recognized by the adoption of prototyping: (i) supporting creativity; (ii) encouraging communication; and (iii) permitting early evaluations. Prototyping models are also useful in the process of capturing system requirements, whether functional or nonfunctional. For example, prototype modeling can be used in the user experience improvement and validation process (Almeida & Monteiro, 2017).

There are several tools and techniques that can be used for building prototypes. We can use, for instance, basic utensils such as paper, pencil, and sticky notes just for paper prototype. Also, we can use software prototyping tools to perform this task. The following software solutions appear to be good options: InVision[1], Marvel[2], Mockplus[3], or Balsamiq[4]. In the context of this book, we will adopt the Balsamiq software for prototyping.

Balsamiq Mockups is a graphical user interface that lets software engineers to create easily mockups and website wireframes. It allows the

---

[1] https://www.invisionapp.com
[2] https://marvelapp.com
[3] https://www.mockplus.com
[4] https://balsamiq.com

creation of pre-built widgets using a drag and drop WYSIWYG editor. The software lets the creation of mockups for desktop apps, mobile apps, websites, dialog windows, Web apps, and tablet apps.

5.2 Scenario I – TourMCard

5.2.1 Prototype Features

The actors of *TourMCard* that will interact with the platform are:

- Tourist – person who visits a given city for leisure purposes for a short period of time;
- Merchant – an entity that owns a local business in a given city. It can be a traditional shop, restaurant, clothing store, etc.;
- Admin – the administrator of the IT platform.

We start by showing all the product features using UML uses cases. Use cases were grouped considering different actors using UML packages. This situation is illustrated in Figure 5.2.

Then, we detail the use cases for each UML package. We start doing it for the tourist actor, as presented in Figure 5.3. The tourist can access the

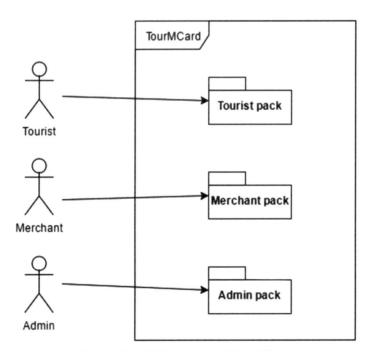

Figure 5.2   UML use cases of TourMCard.

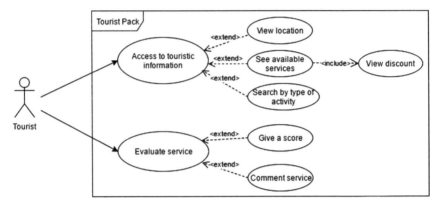

Figure 5.3   UML use cases of Tourist pack.

tourist information of a given city and search all the services by choosing a type of activity. When the tourist checks for an available service, he/she can see the associated discount. Finally, the touristic can evaluate the service by providing a score, a comment, or both.

The use cases that belong to the Merchant pack are presented in Figure 5.4. The merchant can give a discount to a touristy and check the statistics of his/her company. Finally, it can place an advertising, calculate its costs, and perform the corresponding payment online.

Finally, the admin will have access to the use cases presented in Figure 5.5. The admin can manage the credentials of tourist or merchant and provide equipment (e.g., magnetic touristy cards).

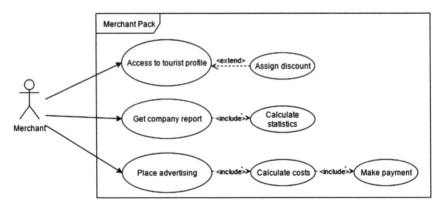

Figure 5.4   UML use cases of Merchant pack.

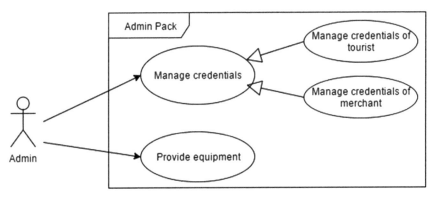

Figure 5.5   UML use cases of Admin pack.

From Tables 5.3 to 5.9 we give a detailed view of each functional requirement.

In order to understand the operating details of the main systems require-ments, we use additional UML diagrams (e.g., activity, sequence, and state).

Table 5.3   Use case: Access to touristic information

ID	TMC_RF1
Name	Access to touristic information
Description	It allows the tourist to see and search for the available services in a given city.
Actor	Tourist
Priority	High
Assumptions	The city where the tourist is located must be automatically detected based on his GPS location of the cell phone.
Pre-conditions	N/A
Post-conditions	N/A

Table 5.4   Evaluate service

ID	TMC_RF2
Name	Evaluate service
Description	The tourist evaluates the service by providing a score (1–5), a comment, or both.
Actor	Tourist
Priority	Moderate
Assumptions	The tourist evaluates the service within 5 days.
Pre-conditions	Service is already available in the database.
Post-conditions	Feedback is given to merchant.

Table 5.5   Access to tourist profile

ID	TMC_RF3
Name	Access to tourist profile
Description	The merchant can access to the tourist profile and assign a customized discount.
Actor	Merchant
Priority	Low
Assumptions	The merchant has an easy interactive system search method to search for a specific tourist.
Pre-conditions	Tourist profile is public.
Post-conditions	Feedback is given to the tourist.

Table 5.6   Get company report

ID	TMC_RF4
Name	Get company report
Description	The merchant can get a complete report of the company's performance with detailed statistics.
Actor	Merchant
Priority	Low
Assumptions	The statistics available in the report can be personalized by the user.
Pre-conditions	Company profile is available in the database.
Post-conditions	N/A

Table 5.7   Place advertising

ID	TMC_RF5
Name	Place advertising
Description	The merchant can buy an ads place. Using the same system, he can estimate the cost and pay it.
Actor	Merchant
Priority	Very low
Assumptions	Different types of ads with various prices are available.
Pre-conditions	Company profile is available in the database.
Post-conditions	Advertising information is available in the merchant profile.

Table 5.8   Manage credentials

ID	TMC_RF6
Name	Manage credentials
Description	The admin can manage the access credentials of tourists and merchants.
Actor	Admin
Priority	Very high
Assumptions	At least one admin exists in the database.
Pre-conditions	N/A
Post-conditions	Send the access credentials to the email account of tourist or merchant.

Table 5.9   Provide equipment

ID	TMC_RF7
Name	Provide equipment
Description	The admin must provide the equipment (smart card) to the tourist.
Actor	Admin
Priority	Very high
Assumptions	Stock of smart cards must exist in warehouse.
Pre-conditions	Tourist must be registered.
Post-conditions	Card is sent to the address of tourist or he can pick it at the tourist point.

The "TMC_RF1" is detailed using an activity diagram that describes the activities that a tourist can do when requesting access to touristy information (see Figure 5.6). Initially, the application needs to detect the city where the tourist is located. After that, the user can choose to get information about the city or search for available services in the city. Finally, detailed information about the service, including available discounts, is provided to the tourist.

The "TMC_RF2" is also detailed using a UML state diagram that provides an overview of the several states of a service (see Figure 5.7). Initially, we start by not having any information about the service. Then, each tourist can comment and/or rate the service. It is not mandatory for the tourist to comment or rate the service.

Finally, we use a sequence diagram to detail the place advertising scenario that corresponds to the functional requirement with id "TMC_RF5" (see Figure 5.8). Initially, the merchant needs to specify the ads type and further information. After that, the system is responsible for estimating the costs and providing this information together with payment details to the merchant. Finally, the payment of the advertising is performed by the merchant.

5.2.2 Technology Choice and Architecture

The physical architecture of the system is depicted in Figure 5.9 using a UML deployment diagram. The application offers a three-tier layer: presentation, business logic, and database. Different technologies are used in each layer. In the presentation layer, we have Android, IoS, and Windows Phone, which allow the application to be used by different cell phones equipment. In the business logic layer, we adopt the .NET framework to create an ASPX application. Finally, the database layer is composed of a database implemented in Microsoft SQL Server 2016.

For the logical architecture of our system, we use seven components and one library, like it is presented in Figure 5.10. The "credentials.aspx"

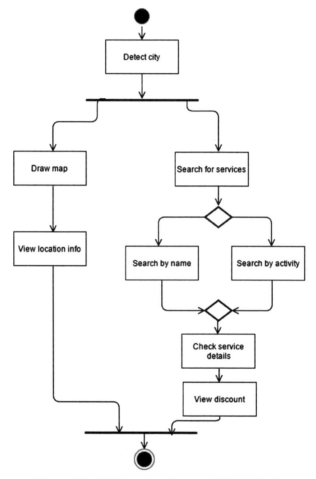

Figure 5.6   UML activity diagram of "TMC_RF1".

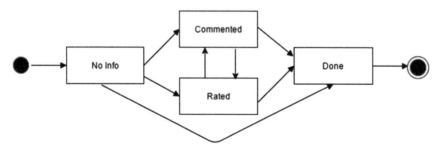

Figure 5.7   UML state diagram of "TMC_RF2".

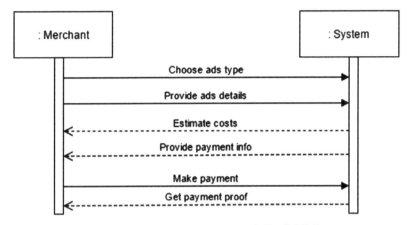

Figure 5.8   UML state diagram of "TMC_RF5".

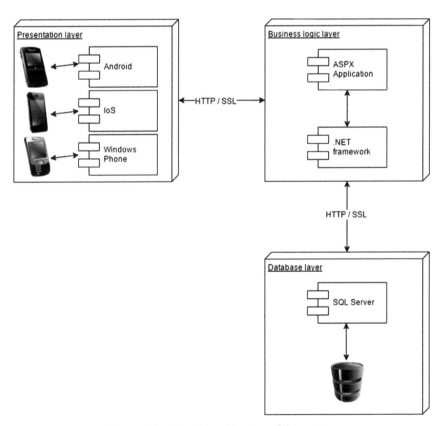

Figure 5.9   Physical architecture of the system.

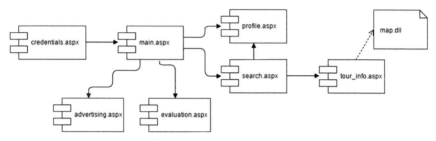

Figure 5.10   Logical architecture of the system.

is responsible for managing all the access permissions to the system, and, after that, the user is forwarded to the "main.aspx" component. There, the user can see his/her profile, search for a profile or touristy information, evaluate a service or place an advertising. Finally, the "tour_info.aspx" uses the "map.dll" library that is responsible for offering external services to access the GPS user's location.

Finally, the class diagram is responsible for presenting the database architecture (see Figure 5.11). The diagram is composed of six classes. The

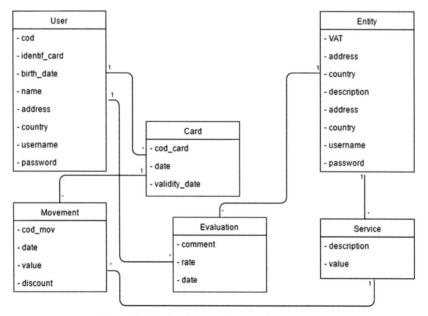

Figure 5.11   Database modeling of TourMCard.

user has an assigned card, and for each card, it is important to record all the performed movements that are always associated with a service. Finally, the user can evaluate an entity by placing a rate or a comment.

5.2.3 Prototype Modeling

The prototype was modeling using the Balsamiq Mockup software. The first screen that appears to the user is the login interface (see Figure 5.12). In this interface, the user only needs to provide his/her username and password. On the end of this screen appears the copyright information about the use of *TourMCard* application.

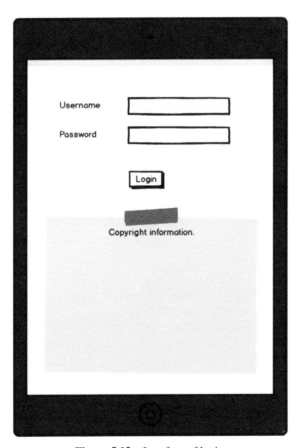

Figure 5.12   Interface of login.

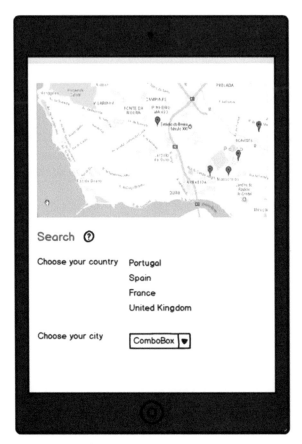

Figure 5.13   Interface of searching for services.

A common operation that a tourist can do is to search for services in a given area. For that, all the available entities are drawn in a navigated and interactive map. The map can be customized for a specific location (see Figure 5.13). For that, the tourist needs to provide information about the country and city.

The merchant can get the company report and calculate the statistics of his company. The statistics show information about the number of visitors, total value of discounts, average rating, and feedback received from visitors (see Figure 5.14). This information is updated in real time.

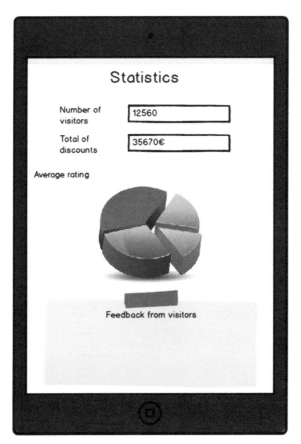

Figure 5.14 Interface of statistics.

Another functionality that the merchant can use is to place an advertising. The merchant needs to specify the type of advertising and its duration. After that, the system estimates the associated costs and the merchant can proceed with the payment using PayPal, MB Net, or credit card (see Figure 5.15).

Finally, another important interface is the evaluation of a service. The user may choose the company name, select the service, and give a score for each characteristic of the service (see Figure 5.16). Additionally, the tourist can leave a comment that will be become visible for the merchant and other users.

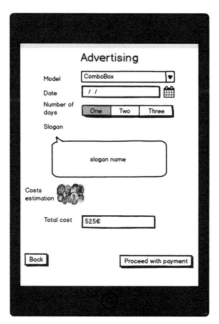

Figure 5.15 Interface of place an advertising.

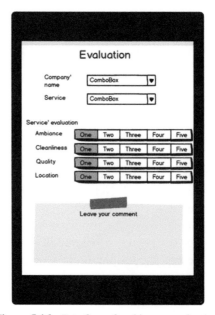

Figure 5.16 Interface of making an evaluation.

5.3 Scenario II – AuditExpert

5.3.1 Prototype Features

The following actors interact with the *AuditExpert* application:

- Client – company that can use the services provided by the application;
- Admin – the administrator of the IT platform;
- Auditor – agent that performs an audit in the IT security network.

The use cases offered by the system are depicted in Figure 5.17. We grouped the use cases using UML packages to facilitate their interpretation.

The client pack is composed of the functionalities presented in Figure 5.18. These features are organized also in packages.

The client pack is composed of four packages. These packages represent the four classes of functionalities provided by the system to the client. We will start by detailing the requirements offered by the "communication and messaging pack" (Figure 5.19). Using this package, the client can access email services, voice services, and fax services.

The "platform and infrastructure pack" is responsible for offering crucial functionalities of the system. The client can manage the server storage and

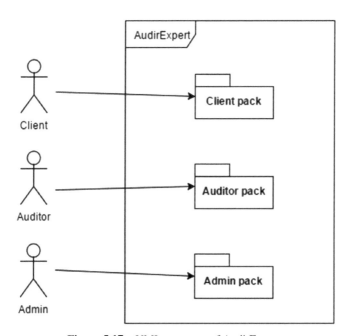

Figure 5.17   UML use cases of AuditExpert.

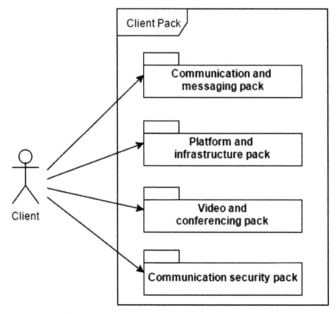

Figure 5.18   UML use cases of client pack.

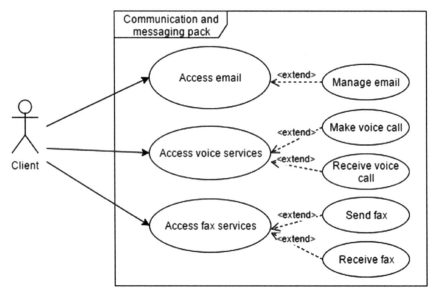

Figure 5.19   UML use cases of communication and messaging pack.

monitor the system performance. Additionally, the client can access virtual machines (VMs), where he can also create new ones. These functionalities are depicted in Figure 5.20.

The "video and conferencing pack" offers synchronous communication features (Figure 5.21). The client can establish a new direct video call or a conference call. It is mandatory to make a new video call always when the

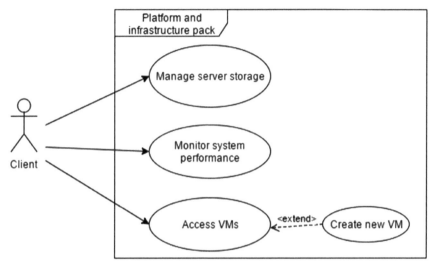

Figure 5.20   UML use cases of platform and infrastructure pack.

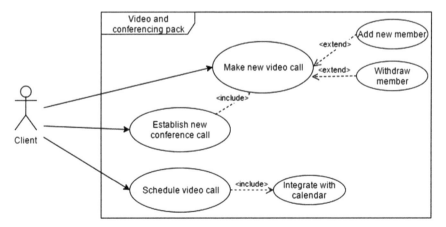

Figure 5.21   UML use cases of video and conferencing pack.

client establishes a new conference call. Finally, the client can schedule a video call that will be integrated with the calendar.

The "communication security pack" offers functionalities in terms of network security (Figure 5.22). The client can manage the firewall, VPN, and IDS. Besides that, the client can establish an anti-virus policy by managing the anti-virus, anti-malware, and anti-spam applications.

The "auditor pack" is responsible for managing all events associated with a security audit (Figure 5.23). The auditor performs an audit in the system. There are three kinds of audits: IT security network, Web applications, and mobile solutions. There is a creation of a report associated with each

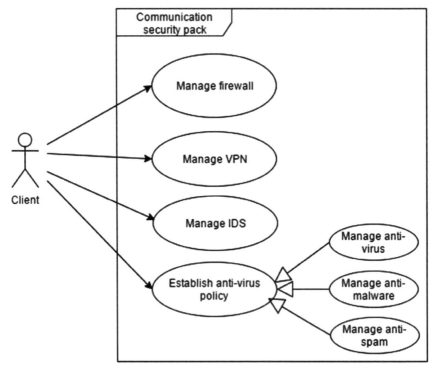

Figure 5.22   UML use cases of communication security pack.

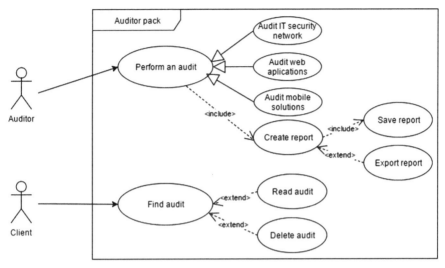

Figure 5.23   UML use cases of auditor pack.

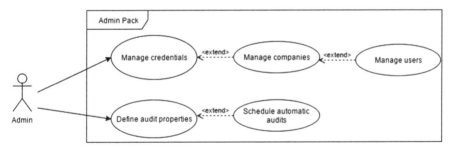

Figure 5.24   UML use cases of admin pack.

performed audit, which can be exported (optional) or saved (mandatory). Finally, the client can find audits already performed by the auditor. The client can read or delete a specific audit.

Finally, the admin is responsible for managing the access credentials to the application (Figure 5.24). For that, the admin can manage companies and users. Additionally, the admin can customize the audit and schedule automatic audits, which will be automatically made by the auditor in a specified date.

The properties of each functional requirements are given from Tables 5.10 to 5.26.

Table 5.10   Use case: Access email

ID	AEXP_RF1
Name	Access email
Description	It allows the client to access and manage all email accounts associated to its company.
Actor	Client
Priority	High
Assumptions	A company may have multiple email accounts.
Pre-conditions	N/A
Post-conditions	Created email accounts may be accessed using webmail or an email client application. Access credentials and configuration script will be sent to the client.

Table 5.11   Use case: Access voice services

ID	AEXP_RF2
Name	Access voice services
Description	It allows the client to use the voice services, including making or receiving a phone call.
Actor	Client
Priority	High
Assumptions	Voice services use Voice over Internet Protocol (VoIP)
Pre-conditions	N/A
Post-conditions	N/A

Table 5.12   Use case: Access fax services

ID	AEXP_RF3
Name	Access fax services
Description	It allows the client to use the fax services, including sending or receiving a fax call.
Actor	Client
Priority	Moderate
Assumptions	Fax services also use VoIP.
Pre-conditions	N/A
Post-conditions	N/A

Table 5.13   Use case: Manage server storage

ID	AEXP_RF4
Name	Manage server storage
Description	It allows the client to manage the capacity of the server.
Actor	Client
Priority	High
Assumptions	Max capacity is 5TB.
Pre-conditions	The server is already created.
Post-conditions	All the functionalities of "communication and messaging pack" can now be used.

Table 5.14   Use case: Monitor system performance

ID	AEXP_RF5
Name	Monitor system performance
Description	It allows the client to monitor the several allocated services.
Actor	Client
Priority	Moderate
Assumptions	N/A
Pre-conditions	The server is already created.
Post-conditions	N/A

Table 5.15   Use case: Access VMs

ID	AEXP_RF6
Name	Access VMs
Description	It allows the client to access virtual machines and create a new one if needed.
Actor	Client
Priority	Moderate
Assumptions	Client can use VirtualBox or VMWare.
Pre-conditions	N/A
Post-conditions	Client can check the status of each virtual machine.

Table 5.16   Use case: Make new video call

ID	AEXP_RF7
Name	Make new video call
Description	It allows the client to make a new video call. Within a video call, the client can add or withdraw a member.
Actor	Client
Priority	High
Assumptions	Video calls also use VoIP services. The client has a camera and its desktop or mobile equipment.
Pre-conditions	N/A
Post-conditions	N/A

Table 5.17   Use case: Establish new conference call

ID	AEXP_RF8
Name	Establish new conference call
Description	It allows the client to define a new conference call with several members.
Actor	Client
Priority	Moderate
Assumptions	N/A
Pre-conditions	Make a new video call must be working.
Post-conditions	N/A

Table 5.18   Use case: Schedule video call

ID	AEXP_RF9
Name	Schedule video call
Description	It allows the client to program a new video call. This event will be integrated with the calendar.
Actor	Client
Priority	Low
Assumptions	The calendar api is properly installed and configured.
Pre-conditions	Make a new video call must be working.
Post-conditions	Event is listed in the calendar.

Table 5.19   Use case: Manage firewall

ID	AEXP_RF10
Name	Manage firewall
Description	It allows the client to manage the firewall.
Actor	Client
Priority	High
Assumptions	N/A
Pre-conditions	The server is properly created and configured.
Post-conditions	N/A

Table 5.20   Use case: Manage VPN

ID	AEXP_RF11
Name	Manage VPN
Description	It allows the client to manage the VPN.
Actor	Client
Priority	High
Assumptions	N/A
Pre-conditions	Firewall is properly configured.
Post-conditions	N/A

Table 5.21 Use case: Manage IDS

ID	AEXP_RF12
Name	Manage IDS
Description	It allows the client to manage the IDS.
Actor	Client
Priority	High
Assumptions	N/A
Pre-conditions	N/A
Post-conditions	N/A

Table 5.22 Use case: Establish anti-virus policy

ID	AEXP_RF13
Name	Establish anti-virus policy
Description	It allows the client to establish an anti-virus policy by managing the anti-virus, anti-malware, and anti-spam applications.
Actor	Client
Priority	High
Assumptions	N/A
Pre-conditions	N/A
Post-conditions	N/A

Table 5.23 Use case: Perform an audit

ID	AEXP_RF14
Name	Perform an audit
Description	The auditor performs an audit in the IT security application, Web application, or mobile solution. The report can be saved or exported to an external application
Actor	Auditor
Priority	Low
Assumptions	Web application is created in PHP or ASPX. Mobile application is created in Android or IoS.
Pre-conditions	N/A
Post-conditions	A report is saved in the application and can be accessed by the client.

Table 5.24 Use case: Find audit

ID	AEXP_RF15
Name	Find audit
Description	It allows the client to search for audits.
Actor	Client
Priority	Low
Assumptions	N/A
Pre-conditions	Audit was initially performed by the auditor.
Post-conditions	N/A

Table 5.25   Use case: Manage credentials

ID	AEXP_RF16
Name	Manage credentials
Description	It allows the administrator to manage the credentials of companies and users.
Actor	Admin
Priority	High
Assumptions	Password data is encrypted using SHA-3.
Pre-conditions	N/A
Post-conditions	N/A

Table 5.26   Use case: Define audit properties

ID	AEXP_RF17
Name	Define audit properties
Description	It allows the administrator to customize the process of an IT security audit. Additionally, automatic IT audits can be scheduled.
Actor	Admin
Priority	Low
Assumptions	Properties of performed audits will remain unchangeable.
Pre-conditions	N/A
Post-conditions	Log information about executed changes in IT audits is automatically stored.

The "AEXP_RF1" requirement demands further detail using an activity diagram (Figure 5.25). The client when accesses the email of his/her company needs to introduce the access credentials. This information needs to be provided earlier by the system administrator. After that, the client has generally four types of available operations: (i) read messages, (ii) compose messages, (iii) manage contacts, and (iv) manage folders. The user needs always to search for a given message before reading a message. On the other hand, when sending a message, it must be indicated if there exists a given attachment. However, it is always possible to send messages both with and without attachments.

The "AEXP_RF8" requirement is also detailed using an activity diagram (Figure 5.26). In order to perform a new conference call, the client needs to define the properties of the conference (e.g., record call, request name, attendee tone, etc.). After that, a new video call is performed. Then, the client establishes a normal conversation with all or specific members. Along the call, a new member can be added or a current member can be withdrawn.

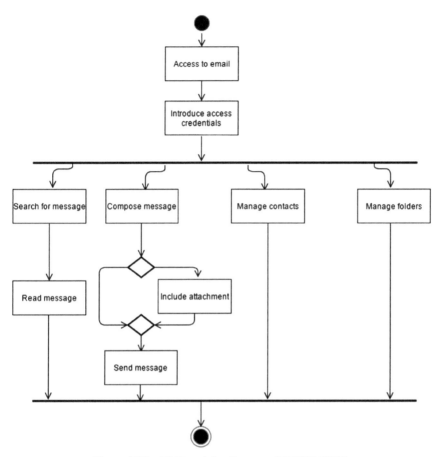

Figure 5.25   UML activity diagram of "AEXP_RF1".

The "AEXP_RF9" requirement is also detailed using a sequence diagram (Figure 5.27). The client can schedule a video call by providing information about data/hour and its members. The system requests the client to validate this information and the data organized in the calendar application after receiving the confirmation from client.

Finally, the "AEXP_RF14" requirement is also detailed using an activity diagram (Figure 5.28). The auditor is responsible for performance and audit

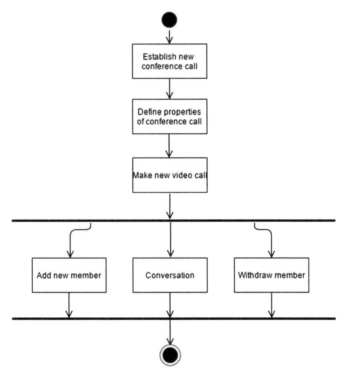

Figure 5.26   UML activity diagram of "AEXP_RF8".

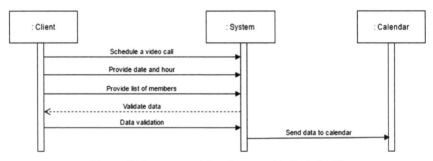

Figure 5.27   UML activity diagram of "AEXP_RF9".

type. For that, he needs to read the audit type object that will inform him if the auditor shall be performed in the IT security network, Web application, or mobile solution. A report is created, and posteriorly saved, after the execution of the audit. Finally, the audit can be exported to a specific format.

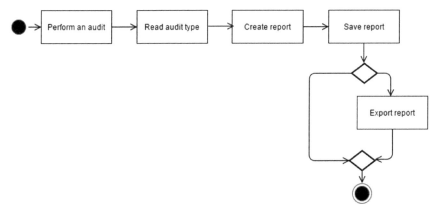

Figure 5.28 UML activity diagram of "AEXP_RF14".

5.3.2 Technology Choice and Architecture

The prototype is built around Microsoft technologies, particularly the C# object-oriented language. The application is composed of three layers: presentation, business logic, and database. In the presentation layer, we use the Windows Presentation Foundation (WPF) that is responsible for rendering the several interface windows for the end-user. In the business logic, we adopt the C# language and the XAML as a declarative markup language used to define the application's user interface. Finally, the database layer offers access to a Microsoft SQL Server 2016 database. The physical architecture is depicted in Figure 5.29.

The logical architecture is composed of 11 components and 2 libraries, as shown in Figure 5.30. The "credentials.cs" is responsible for managing all the accesses to the system from the client and administrator. The "main.cs" is the class responsible for accessing all the elements of the system. The "video_conf.cs" class needs to access the "calendar.dll" library to schedule a video conference call. The "audit.cs" is responsible for managing the operations related to the realization of IT audits and it uses two depen-dence classes: "find.cs" and "export.cs". Finally, the "admin.cs" manages the credential accesses and uses the "audit_config.cs" class to customize the IT audits. An automatic IT audit needs to be included in the windows registry by calling the "registry.dll" library.

Finally, the class diagram of the system is presented in Figure 5.31. The database is composed of nine main classes and one derivative class that result from the many-to-many relationship between the "client" and

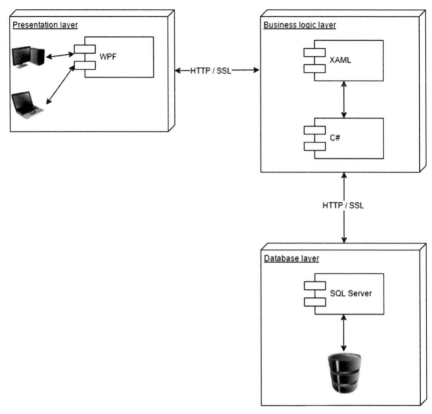

Figure 5.29   Physical architecture of the system.

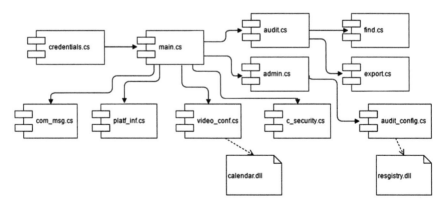

Figure 5.30   Logical architecture of the system.

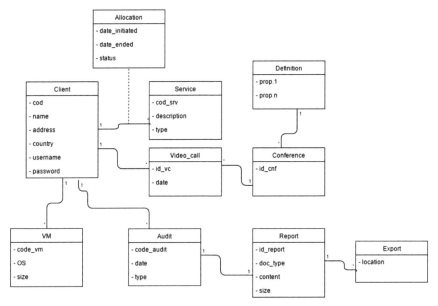

Figure 5.31 Database modeling of TourMCard.

"service" classes. The type of service can be a communication service, a platform infrastructure, or a communication security artifact. An audit is also associated to one client. Each audit has only one report that can be exported multiple times. It is also important to record information about the location where the report was exported.

5.3.3 Prototype Modeling

The prototype was modeled using the Microsoft Visual Studio 2015. The first screen is the typical authentication window that requests the user to enter his/her access credentials. This situation is illustrated in Figure 5.32.

Then, the user accesses the dashboard. The dashboard is composed of a menu that offers a wide range of features in terms of communication services, infrastructure services, and security applications. The interface of the dashboard is depicted in Figure 5.33.

Inside each menu, the client accesses specific services inside each segment. The organization is done by adopting a sub-menu navigation technique, which is illustrated in Figure 5.34.

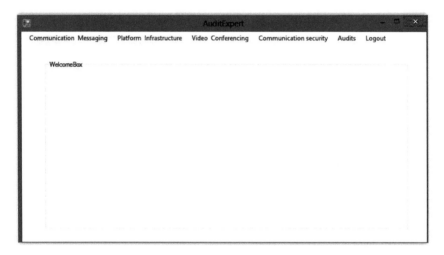

Figure 5.32   Interface of login.

Figure 5.33   Interface of dashboard.

An important feature is offered by the "Audits" menu entrance (Figure 5.35). Using it, the client can search for a specific audit by date or type. The list of available audits is dynamically updated according to the filter criteria. The audit content is shown in the right part of the window.

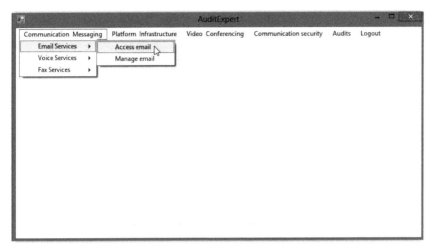

Figure 5.34   Interface of dashboard (organization by sub-menus).

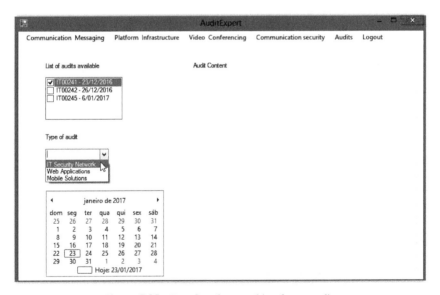

Figure 5.35   Interface for searching for an audit.

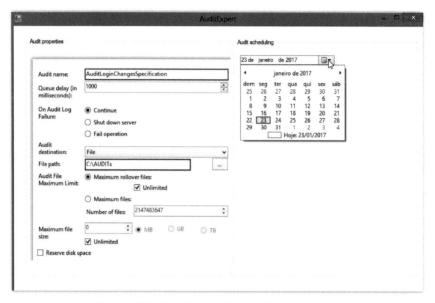

Figure 5.36   Interface for the definition of an audit.

Finally, it is important to show how the administrator can configure an audit (Figure 5.36). For that, in the audit properties section (left side of the window), the user can specify the audit name, queue delay, audit log, and destination and define a maximum limit for the audit. In the audit scheduling section (right side of the window), the user can define a schedule for the execution of the audit.

Conclusions and Future Research Topics

Conclusions

The importance of entrepreneurship is unquestionable, as it contributes to the creation of jobs, the promotion of creativity and innovation, and the development of a country's economy and society. Being an entrepreneur is increasingly a professional option for young graduates and for people already with a considerable professional career. Thus, promoting and valuing entrepreneurship and an entrepreneurial culture in a society is vital for social, economic, technological, and organizational development. In this way, the implementation of a strategy focused on entrepreneurship education becomes fundamental for changing attitudes and behaviors.

In the process of establishing a start-up, it becomes critical to create a business plan. This book is a powerful tool for businesses and entrepreneurs. In this book, we have documented the whole process of growth of a company until its maturity. The business plan describes human, technological, and financial resources, observing key points, identifying opportunities, and anticipating potential difficulties. This tool is vital to the success of start-ups, even for those who do not seek any type of financing or capitalization.

The business plan template that we suggest is composed of five chapters plus an executive summary. In each of these sections, we have the following objectives:

- Executive summary – summarizes the main elements of the business plan and the decisions taken by the entrepreneurs. Therefore, the objectives of the business must be presented, justifying the market need, growth potential, differentiated aspects of the business, necessary resources, and financial analysis;
- Chapter 1: Business concepts – introduces the idea of business, mission, vision, value proposition, company organization, innovation policy, human resources, business risks, and legal considerations;
- Chapter 2: Marketing plan – describes the marketing policy followed by the company, beginning with the presentation of the work of

marketing research, analysis of the environment, customers, partners, and competitors. Additionally, it is important to present the company's strategic positioning, advertising and sales strategies, and marketing mix. Finally, the business model canvas is also presented in this chapter;

- Chapter 3: Operational plan – describes the production process, the relationship with suppliers, inventory, payment policy, and quality management policy. The presentation of the action plan is also highlighted in this chapter with the most important key dates throughout the development of the business;
- Chapter 4: Financial plan and viability analysis – the financial plan is presented with its feasibility analysis. For this purpose, the initial investment required and the cash flow approach for the first 6 years of the business execution are detailed. Finally, we calculate the main indicators such as NPV, IRR, and payback;
- Chapter 5: Prototype description – the prototype developed in the IT component is presented. For this purpose, it presents the functional and non-functional requirements of the project, the technological architecture that supports the application and a UML modeling of the main requirements of the project.

It is also important to highlight that the development of the business plan was guided by the use of methodologies and bibliographical references, contributing to all phases of research, strategy, and analysis of the company, which makes it essential for the elaboration of the business plan.

Managerial Implications

This book aims to help the process of designing, launching, and accompanying a technological start-up in the IT field. It aims to be a reference guide in documenting the strategic and tactical decisions of an IT project for all involved stakeholders. In the same way, it is also intended to help entrepreneurs in capturing new investors and financing, either through business angels, venture capital, or banking institutions.

Academic Implications

The main objective of this book is to be a reference manual for higher education institutions that offer courses or curricular units in the entrepreneurship field. It is intended that this book helps students to understand the fundamental concepts in the area of management and application of software engineering knowledge to the entrepreneurship field. This book can still be used as a reference manual in the follow-up of an IT project to write a business plan.

Future Research

The process of launching a new IT business is an extremely challenging task due to the constant technological developments and changes in market needs. Increasingly technologies are immersive and are part of our daily lives. Accompanying the technological evolution and analyzing how these new technologies inhibit the potential of emergence of new business models is an area that we certainly want to explore. Associated with this, we also have the emergence of new theories of strategic positioning and innovation strategies whose inclusion in business plans becomes relevant. It is also important to recognize the role of science parks in leveraging new business in the IT arena and, consequently, it becomes useful to explore the impact of science parks on these businesses.

One of the most innovative components of this business model template is the inclusion of "Chapter 5 – Prototype Description", in which the IT component of the project is presented. It is important to explore how this component is important for IT companies at two levels: (i) in monitoring project implementation and (ii) in the exploitation of financing sources.

Finally, we also want to gather feedback from entrepreneurs about the proposed business plan template. With this objective in mind, we are willing to improve all the components of the business plan and make this model adaptable and evolve according to the dynamics of the market.

Bibliography

AEC, RI, KUMULUR, & TNAU (2012). Management Approaches. PO: 621712. Available at: http://pt.slideshare.net/SarbojeetJana/management-approach-swot

Almeida, F., Silva, P., & Leite, J. (2017). Proposal of a carsharing system to improve urban mobility. *Theoretical Empirical researches in Urban Management (TERUM)*, *12*(3), 32–44.

Almeida, F., & Monteiro, J. (2017). Approaches and Principles for UX Web Experiences. *International Journal of Information Technology and Web Engineering*, *12*(2), 49–65.

Banerjee, A. (2014). What a Prototype Is (and Is Not). UX Magazine, Article No. 1345. Available at: https://uxmag.com/articles/what-a-prototype-is-and-is-not

Beaudouin-Lafon M, & Mackay W. (2003). Prototyping tools and techniques. In: Sears A, Jacko JA, editors. *The Human-Computer Interaction Handbook*. Mahwah: Lawrence Erlbaum Associates.

Bierman, H. (2007). Comparing Net Present Value and Internal Rate of Return. Available at: http://www.financepractitioner.com/cash-flow-management-best-practice/comparing-net-present-value-and-internal-rate-of-return?full

Dalbey, J. (2016). Non-functional Requirements. Available at: http://users.csc.calpoly.edu/jdalbey/SWE/QA/nonfunctional.html

Damodaran, A. (2007). Probabilistic Approaches: Scenario Analysis, Decision Trees and Simulations. Available at: http://www.stern.nyu.edu/~adamodar/pdfiles/papers/probabilistic.pdf

Drake, P. (2016). Financial Ratio Analysis. Available at: http://educ.jmu.edu/~drakepp/principles/module2/fin_rat.pdf

Ehmke, C., Fulton, J., Lusk, J. (2015). Marketing's Four P's: First Steps for New Entrepreneurs. Purdue University, EC-730. Available at: https://www.extension.purdue.edu/extmedia/ec/ec-730.pdf

Felici, M. (2007). Deployment Diagrams. University of Edinburgh. Available at: http://www.inf.ed.ac.uk/teaching/courses/seoc/2007_2008/notes/LectureNote14_notes.pdf

Fine, L. (2009). *The SWOT Analysis: Using your Strength to overcome Weaknesses, Using Opportunities to overcome Threats*. New York: CreateSpace Independent Publishing Platform.

Gerth, D. (2016). Unit 13: Channels of Distribution, Logistics, and Wholesaling. Available at: http://ww2.nscc.edu/gerth_d/MKT2220000/Lecture_Notes/unit13.htm

Goi, C. (2009). A Review of Marketing Mix: 4PS or More?. *International Journal of Marketing Studies*, *1*(1), 2–15.

Gupta, A. (2013). Environmental and PEST analysis: An approach to external business environment. *Merit Research Journal of Art, Social Science and Humanities*, *1*(2), 13–17.

IAA (2013). Stress Testing and Scenario Analysis. International Actuarial Association. Available at: http://www.actuaries.org/CTTEES_SOLV/Documents/StressTestingPaper.pdf

IDC (2012). Worldwide Security and Vulnerability Management 2013–2017 Forecast and 2012 Vendor Shares. IDC Consulting Services. Available at: https://www.qualys.com/docs/idc-worldwide-vulnerability-management-2013–2017-forecast-2012-vendor-shares.pdf

Info Entrepreneurs (2016). Identify and Sell more to Your Most Valuable Customers. Available at: http://www.infoentrepreneurs.org/en/guides/identify-and-sell-more-to-your-most-valuable-customers/

Info Entrepreneurs (2017). Stock Control and Inventory. Available at: http://www.infoentrepreneurs.org/en/guides/stock-control-and-inventory/

Ingram, D. (2016). Financial Business Objectives. Available at: http://small business.chron.com/financial-business-objectives-4072.html

Iniazz, J., & Popovich, K. (2003). Understanding customer relationship management (CRM): people, process and technology. *Business Process Management Journal*, *9*(5), 672–688.

ISO/IEC 27001 (2017). ISO/IEC 27001 – Information security management. Available at: http://www.iso.org/iso/iso27001

ISO 9001:2008 (2017). ISO 9001:2008 – Quality management systems – Requirements. Available at: http://www.iso.org/iso/catalogue_detail?cs number=46486

Ittelson, T. (2009). Financial Statements: *A Step-by-Step Guide to Understanding and Creating Financial Reports*. New Jersey: Career Pr Inc.

Kent, R. (2006). *Marketing Research: Approaches, Methods and Applications in Europe*. Andover: Cengage Learning EMEA.

Kim, W., & Mauborgne, R. (2015). *Blue Ocean Strategy, Expanded Edition: How to Create Uncontested Market Space and Make the Competition Irrelevant*. Brighton: Harvard Business Review Press.

KPMG (2013). Travel and tourism sector: Potential, opportunities and enabling framework for sustainable growth. KPMG Company. Available at: http://www.kpmg.com/IN/en/IssuesAndInsights/ArticlesPublications/Documents/KPMG-CII-Travel-Tourism-sector-Report.pdf

Kurkovsky, S. (2012). Distributed Systems Architectures. Central Connecticut State University. Available at: http://www.cs.ccsu.edu/~stan/classes/cs530/slides/se-12.pdf

Lowe, C. & Zemliansky, P. (2011). *Writing Spaces 2*. West Lafayette: Parlor Press.

Mohammadhossein, N., & Zakaria, N. (2012). CRM Benefits for Customers: Literature Review. *International Journal of Engineering Research and Applications*, 2(6), 1578–1586.

Montes, M. (2013). Analytic tools to move from a Red Ocean Strategy to a Blue Ocean Strategy (part 1). Available at: http://www.an-entrepreneur.com/public-content/marketing/analytic-tools-to-move-from-a-red-ocean-strategy-to-a-blue-ocean-strategy-part-1/

MSG (2016). Financial Planning – Definition, Objectives and Importance. Available at: http://www.managementstudyguide.com/financial-planning.htm

OECD (2006). Successful partnerships a guide. Forum on Partnerships and Local Governance, Centre for Social Innovation (ZSI). Available at: http://www.oecd.org/leed-forum/publications/FPLG_Guide_2006.pdf

OECD (2016). Enterprises by business size. Available at: https://data.oecd.org/entrepreneur/enterprises-by-business-size.htm

Osterwalder, A., & Pigneur, Y. (2010). *Business Model Generation: A Handbook for Visionaries, Game Changers, and Challengers*, New Jersey: Wiley.

Porter, M. (1998). *Competitive Strategy: Techniques for Analyzing Industries and Competitors*. New York: Free Press.

Porter, M. (2008). The Five Competitive Forces that Shape Strategy. *Harvard Business Review*, No. 1, pp. 24–41.

Proctor, T. & Jamieson, B. (2004). *Marketing Research*. New Jersey: Pearson Education.

Rasmusson, J. (2017). Agile in a Nutshell. Available at: http://www.agilenutshell.com/scrum

Reifer, D., & Boehm, B. (2006). *Software Management*. New York: Wiley-IEEE Computer Society.

Rosenbloom, B. (2011). *Marketing Channels*. Nashville: South-Western College Pub.

Royce, W. (1998). *Software Project Management: A Unified Framework*. New Jersey: Pearson Education.

Sánchez, A. (2016). Theme 8: Generic Strategies. Available at: http://www.uhu.es/45122/temas/Theme8_presentation.pdf

Shankar, V., & Carpenter, G. (2012). *Handbook of Marketing Strategy*. Cheltenham Glos: Edward Elgar Publishing.

Shukla, P. (2008). *Essentials of Marketing Research*. London: Bookboon.

Sommerville, I. (2015). *Software Engineering*. London: Pearson.

Stan, L., & Nedelcu, A. (2015). Entrepreneurial Skills, SWOT Analysis and Diagnosis in Business Activities. *Review of the Air Force Academy*, *1*(28). 187–190.

Surbhi, S. (2015). Difference Between Income Statement and Cash Flow Statement. Available at: http://keydifferences.com/difference-between-income-statement-and-cash-flow-statement.html

Szopa, P., & Pekala, W. (2012). Distribution channels and their roles in the enterprise. *Polish Journal of Management Studies*, *Vol. 6,* 143–150.

Tanwar, R. (2013). Porter's Generic Competitive Strategies. *IOSR Journal of Business and Management*, *15*(1), 11–17.

TechNavio Insights (2014). Global Cloud Security Software Market. Available at: http://www.trendmicro.com/cloud-content/us/pdfs/business/reports/rpt_technavio-global-security-software-market.pdf

Tutorials Point (2016). UML – Standard Diagrams. Available at: https://www.tutorialspoint.com/uml/uml_standard_diagrams.htm

UML (2016). What is UML?. Available at: www.uml.org/what-is-uml.htm

Webster, A. (2010). Difference Between Income Statement vs. Balance Sheet vs. Cash Flow. Available at: http://thefinancebase.com/difference-between-income-statement-vs-balance-sheet-vs-cash-flow-2064.html

WEC (2013). The Travel & Tourism Competitiveness Report 2013. World Economic Forum. Available at: http://www3.weforum.org/docs/WEF_TT_Competitiveness_Report_2013.pdf

WTTC (2015). Travel & Tourism Economic Impact 2015 WORLD. World Travel & Tourism Council. Available at: http://zh.wttc.org/-/media/files/reports/economic-impact-research/regional-2015/world2015.pdf

Yasanallah, P., & Vahid, B. (2012). Studying the Status of Marketing Mix (7Ps) in Consumer Cooperatives at Ilam Province from Members' Perspectives. *American Journal of Industrial and Business Management*, *2*, 194–199.

Annex I – Porter Five Forces

Overview

The analysis of Porter Five Forces is one of the most used exercises in market segmentation in the process of opening a new business. This analysis has as essential objective of evaluating the attractiveness of each specific segment. From the Porter Five Forces, entrepreneurs have more rigorous and complete information to make strategic decisions, identifying inherent risks of a particular segment, such as dependence on a single supplier, the existence of oligopolies, and the competition between competitors and the emergence of new competitors in the market.

Structure

The template offers five blocks to perform a Porter Five Forces. The following questions should be placed for each block.

Bargaining Power of Buyers

What is the proportion of customers vs. the number of suppliers?

What is the power of customers to establish the rules and terms of business?

How difficult is to attract new customers?

What is the power of customers to affect the opinion of others?

Bargaining Power of Suppliers

How many suppliers are available in the industry?

What is the difference between the suppliers?

What is the cost of moving to a new supplier?

What is the relative power of the company vs. supplier?

Threat of New Entrants

What is the initial cost of opening a new business in my segment?

Are there barriers to the entry of new competitors?

Are there tax incentives available to anyone who wants to enter the industry?

Threat of Substitute of Products/Services

Are there in the market any substitute products or services?

Is there any part of the company's work process that can be automated or replaced?

Is it easy to find alternatives to the solution offered by the company?

Rivalry among Existing Competitors

How many competitors are there in the market?

What are the elements of competition among competitors?

What is the market growth potential?

What is the risk of market decline?

Annex II – SWOT Analysis

Overview

SWOT analysis is one of the best-known and most widely used methods in the field of management, which enables an entrepreneur to find relevant information useful for business decision-making. The structure is composed of four fundamental blocks in the internal perspective (strengths and weaknesses) and external perspective (opportunities and threats). From the SWOT analysis, the entrepreneur is able to make decisions more safely, to have a greater understanding of the organization's functioning, to know the company's position in relation to the competition, to establish alternatives for actions, and to prepare strategies for solving problems.

Structure

The template offers two independent sheets for internal and external analysis. Several elements are considered for each dimension that must be evaluated considering two perspectives: impact and importance. Impact can be classified into five levels (from major threat to major opportunity), and importance offers three levels (from low to medium and high).

External Factors

Political-Legal Aspects

Political situation – political instability causes distrust and insecurity among investors, thus constituting a potential threat.

Sectoral policies – policies that protect certain sectors that facilitate investments can provide opportunities.

Legislation – legal changes may restrict economic activity, create declining demand, or act in the opposite direction.

Sectoral regulations – highly regulated sectors tend to create difficulties in establishing new businesses, which limit possible expansions.

Economic Aspects

The GNP growth rate or industrial production growth rate – economic growth leads to increased demand from existing consumers or the emergence of new ones.

Employment rate – if the employment rate is high, then it can be difficult for the company to be able to find employees and, therefore, it can be considered a threat. On the other hand, low employment rate can facilitate the recruitment process.

Inflation – a very high or low inflation rate can have a great influence on the prices charged and on their updating process.

Interest rate – the existing rate may tend to favor consumption, which is attractive for the existence of loans.

Incentive programs – analysis of the existence of initiatives aimed at dynamizing the economy.

Social, Demographic, and Cultural Aspects

Demographic trends – should consider elements such as birth rate, age pyramid, and trends in the movement of consumers, such as from village to city or from the interior to the coast.

Attitudes of buyers – the attitude that consumers have about the marketed products may be positive or negative.

Lifestyle – the lifestyle that people have can favor the sale of the product.

Technological Aspects

Technological evolution – expectations of the emergence of new manufacturing technologies, or the appearance of alternative products based on technological evolution.

Technological innovation – frequency, ease of appearance of innovations in products, processes, etc.

Time of obsolescence – average duration of the adopted technology.

Alternative technologies – existence of alternative technologies in the production process.

Competition

Competitor image force – what is the perception of the consumers about my main competitors.

Financial capacity – dimension of the financial resources that competitors have.

Ability to react – level of the reaction capacity offered by competition to initiatives developed by our company.

Products – analysis of the portfolio of products offered by the main competitors, such as range and product life cycle.

Price – comparing the prices charged by competitors and their perception by consumers.

Pre-sales service – services provided by competitors prior to sale.

After-sales service – how competitors act in the after-sales service: response time, responsiveness, problem-solving, etc.

Distribution – analysis of dimension, coverage, and services provided.

Communication – diversity, intensity, coverage, adopted channels, etc.

Level of customer satisfaction – analysis of the level of satisfaction of customers who buy products in the competition and exploration of the possibility of changing.

Suppliers

Negotiating skills – capacity of negotiation of our suppliers in the established prices, in terms of payments, granted discounts, etc.

Responsiveness – supply capacity, average supply time, supply failures, etc.

Pre-sales service – services provided by suppliers before the sale and how they can help to improve our business.

After-sales service – quality of after-sales services and how they affect the performance of our company.

Upstream integration – possibility of our suppliers becoming competitors.

Downstream integration – possibility of our suppliers acquiring companies that were their suppliers and affect our business, for example, decreasing the established prices.

Internal Factors

Marketing: Product

Performance – it is associated with the result that is obtained by the use or consumption of the product.

Packaging – analysis of the packaging at the design level, used material, size, communication through the packaging.

Brand image – perception of the customers about a brand.

Portfolio – the collection of items that a company sells.

Range – a set of variations of the same product platform that appeal to different market segments.

Development of new products – ability to develop a new product's line.

Innovation capacity – competence that the company presents in innovating products, services, or processes.

Marketing: Price

Commercial margin – analysis and comparison with the practiced prices offered by competitors.

Discount policy – existence of discounts, practiced prices, frequency, recipients, etc.

Funding policies – financing provided by competitors.

Promotional policy – analysis of the developed promotional actions, typology, return, frequency, objectives, etc.

Elasticity – behavior of the variation of the demand considering the price variation.

Price versus value perceived – level of adjustment between the price practiced by the company and the price perceived by the customers.

Marketing: Placement

Coverage rate – number and location of points of sale. Analysis of the concentration or dispersion of points of sale.

Organization of the channels – how the channels are structured to achieve the objectives.

Control of channels – ability of the company to enable the distribution channels to perform their function according to the established guidelines.

Possible conflicts between channels – existence of possible conflicts between existing channels.

Image distributors – the contribution that the distributors give to the brand image.

Points of sale – developed functions and how they contribute to the increase of brand awareness; analysis of the typology of points of sale.

Marketing: Communication

Reputation of the company – level of positive recognition of the company brand by the purchasers or other relevant audience.

Product notoriety – level of notoriety. It can be spontaneous or assisted.

Advertising – usage level and ads quality.

Public relations – tools and techniques used to create and maintain a positive publicity.

Merchandising – developed promotional actions and available materials at points of sale.

Direct marketing – level of use of this communication model with the client.

Magazines/brochures – existence and quality of this type of material used in marketing communication.

Marketing: Sales Force

Number of sellers – adequacy of the number of existing vendors to the needs.

Degree of motivation – analysis of the sales force's motivational ability.

Rotation of the sellers – stability of the sales force; analysis of the number of exits and entries of new employees.

Performance – results that the sales force is achieving and its comparison with the proposed objectives.

Training in sales techniques – training of employees in sales techniques.

Technical development in products and services – level of knowledge of the marketed products and services, as well as of the organization itself.

Marketing: Sales

Sales volume – comparison of the sales volume that the company is carrying out with the defined objectives.

Sales evolution – analysis of the trend of evolution (decrease, stagnation, growth) of sales over time.

Concentration of turnover (Pareto law) – dependence of the company on a small group of clients.

Market share – value held by the company and comparison with the outlined objectives. It is calculated by dividing the value of the sales of the company by the sales of the sector in analysis or of a determined range of products. Can be calculated based on sales in physical units or monetary value.

Relative market share – value held by the company and comparison with the objectives outlined. It is calculated by dividing the value of the sales of the company by the sales of the main competitor. Can be calculated based on sales in physical units or monetary value. The value is always given in percentage.

Cross-selling – ability of the company to sell different products to different customers than they usually buy.

Up-selling – ability to get customers to buy more sophisticated products.

Marketing: Customer Relationship

Loyalty policy – performed activities developed by the company to have loyal customers.

Fulfillment of commitments – ability of the company to keep promises and established commitments.

Degree of defection – percentage of customers that cease to be.

Customer value – value that customers tend to present over the relationship.

Customer equity – value of the customer's portfolio.

Personalization and customization – analysis of the company's ability to customize the relationship through the customer profile and customize the products, services to meet the specific needs of each customer.

Marketing: Digital

Website – structure of the site, suitability to objectives, usability experience, and responsiveness to the mobile environment.

Blog – level of blog use by the company and analysis of metrics, such as number of followers, number of likes, etc.

SEO – techniques used by the company that influences the position that the company occupies in search engines ranking.

Back links – number of links on external pages that link to company pages.

Social network – level of use of social networks by the company and analysis of metrics, such as engagement rate, number of followers, etc.

Google AdWord – level of use by the company and analysis of the return on investment.

Display advertising – use of online ads.

Ads social: level of use of ads by the company in social networks and associated metrics, such as engagement rate, number of conversions, sales value, etc.

Email marketing – level of use by the company, potentialities of the adopted application, sending newsletters, and associated metrics, such as number of views, number of clicks, etc.

Landing pages – level of use by the company and conversion rate.

Apps development – existence of developed apps by the company to be used, for example, in smartphones, achieving the objectives inherent to its development.

Service and Support

Warranties – duration time, form of activation.

Returns of products – product return policy, particularly in faulty or incorrectly sized goods.

Replacement material – possibility and facilities that the company makes available to the customer to exchange products that are under warranty, or to be offered an alternative when it is not possible to intervene in the damaged product.

Speed and efficiency in the handling of complaints – ability to handle complaints in a way that does not detract the company's image.

Respect of the deadlines – ability of the company to meet the defined deadlines in the area of service and support.

Quality of service – realization that the received service meets the expectations.

Image of pre-sale service – how customers face the service provided in the pre-sale.

Image of after-sales service – how customers face the service provided after sale.

Finance

Net results – amount of income or revenue above and beyond the costs or expenses a company incurs. Analysis of the achieved value and its comparison with the defined objectives and with the competition.

Results/sales – ratio between the obtained results by the costs of sales. The higher is better.

Economic profitability – difference between the revenue a firm earns from sales and the firm's total opportunity costs.

Financial profitability – determine the scope of a company's profit in relation to the size of the business.

Average receipts – activity indicator that seeks to measure the efficiency level with which the company is managing its clients' credit.

Working capital requirement – a financial metric that represents operating liquidity available to a business. Working capital = current assets − current liabilities.

Solvency level – the ability of a company to meet its long-term financial obligations.

Production

Technology level – existence of automation, contribution to production, economies of scale (if relevant).

Compliments of manufacturing deadlines – ability of the company to produce the products in the estimated time agreed with the customer.

Production costs – ability of the company to control the production costs, which will have impact on the establishment of the product's price.

Monetization of equipment – ability of the company to monetize the equipment, avoiding spare times and therefore optimizing the fixed costs.

Quality control – level of quality control and existence of quality certificates.

Production flexibility – ability of the company to adapt its production line to new demands and new products.

Manufacturing waste – ability to optimize the raw material and production process in order to avoid waste.

Human Resources

Management style – leadership style, analysis of internal communication, organizational culture, etc.

Number of employees – analysis of the total number of employees and their distribution by functional areas is appropriated attending the organizational goals.

Recruitment and selection policy – analysis of the HR policies considering the company's strategic goals.

Degree of training – analysis of the company's capacity to assure the formation of its employees.

Remuneration policy – the adopted remuneration format, and the level of remuneration compared to competitors in the sector.

Motivation – level of motivation that the employees show and how they are reflected in the initiative capacity.

Fitness to the role played – analysis of knowledge and professional and personal skills compared with the respective job description.

Annex III – Industry Strategy Canvas

Overview

The industry strategy canvas allows the new venture to position itself in the market taking into account the characteristics of its offer of products or services as well as its main competitors. For this purpose, it is necessary for the entrepreneur to faithfully judge the current state of the company and to establish a different path from the other main competitors.

Structure

The template gives a 2D matrix that must be filled by the entrepreneur. The lines of the matrix represent a competitive factor. A competitive factor is a relevant characteristic of the product or service that is perceived by customers. Some examples of competitive factors are:

- Price;
- Product features;
- Partnerships;
- Advertising;
- Customer relationship;
- Technical support; and
- Quality of the product/service.

In the columns of the matrix, the entrepreneur's company and its main competitors must be represented. For ease of analysis, only the top competitors (e.g., up to three) should be represented. The entrepreneur must make a comparative analysis of each competitive factor for each company using a scale between 1 and 5. A high value should be given when the company shows a high performance in that competitive factor, whereas a low value represents exactly the opposite. Intermediate values may be given, but only integer values can be considered.

Annex IV – Business Model Canvas

Overview

The business model canvas offers a diagram that allows an entrepreneur to schematically visualize the various dimensions of a business structure. This tool becomes very useful for entrepreneurs since it provides a graphical way to achieve an overview of the strategic fit between the different areas of the company.

Structure

The template gives a structure of the nine blocks suggested by Alexander Osterwalder. The following elements should be considered for each block.

Key Partners

Describe the main partners that will help the company to offer its value proposition.

Key Activities

Enumeration of the key tasks that will be performed to build the value proposition.

Key Resources

Enumeration of the key resources, which are critical to assure that all key activities will be performed by the company.

Value Propositions

Presentation of the solutions and benefits brought by the company's offer considering the market needs.

Customer Relationships

Description of the process how the company relates and interacts with its customers.

Channels

Description of the channels that will be used to establish communication with the customer.

Customer Segments

Description of the target audience profile based on their essential characteristics, behaviors, and preferences.

Cost Structure

Cataloging the main costs inherent in the activities of the company, the production process and service to customers.

Revenue Streams

Presentation of the main revenues of the company considering the sale of products and services provided by the company.

Annex V – Financial Plan

Overview

The success of a new venture depends greatly on the entrepreneur's ability to financially manage the business. In fact, the correct projection and systematic monitoring of finances allows taking right decisions. In this sense, the financial plan should help an entrepreneur to evaluate the viability of the business as well as reflect in numbers everything that was written in the various sections of the business plan.

Structure

The template is divided into two sheets: (i) start-up costs and (ii) cash flows. The former is responsible for defining the necessary financing for the opening of a new business, considering internal and/or external financing funds. The latter is responsible for making a projection of business expenses and revenues over several years. This template considers financial projections up to 5 years of activity.

Start-up Costs

Fixed Assets

Under this heading, the fixed expenses needed to set up the business should be considered, namely real estate, buildings, leasehold improvements, equipment, furniture, and vehicles.

Operating Capital

All businesses need start-up capital to meet the company's operating expenses in the initial phase of its activity. It includes expenses such as pre-opening salaries and wages, prepaid insurance premiums, beginning inventory, legal and accounting fees, rent deposits, utility deposits, supplies, advertising and promotions, licenses, and working capital.

Sources of Funding

Internal or external financing should be considered. These elements are presented in percentage terms. There are owner's cash injection, outside investors, and additional loans or debts.

Cash Flows

Incomes

Incomes that the company gets from its activities, including product sales, services, and other incomes. Incomes from both national and international markets should be considered.

Costs

Costs that the company necessarily has from its activities. Among them are investments, loan payments, buildings and renting, salaries, equipments, advertising, supplies, travel, vehicles, telephone and communications, and miscellaneous services.

Financial Indicators

The following indicators are calculated: NPV, IRR, and payback.

Annual Growth Rate

Percentage annual growth rate is defined for the following items: product sales, services, advertising, and supplies.

Index

About the Authors

Fernando Almeida has a PhD in Computer Science Engineering from Faculty of Engineering of University of Porto (FEUP). He also holds an MSc in Innovation and Entrepreneurship and an MSc in Informatics Engineering from FEUP. He has around 10 years of teaching experience at higher education levels in the field of computer science and management. He has also worked for 15 years in several positions as software engineer and project manager for large organizations and research centers like Critical Software, CICA/SEF, INESC TEC, and ISR Porto. During that time, he had the chance to work in partnership with big international organizations and universities in several European projects. He has more than 90 scientific publications on the field of computer science, management, and innovation mainly in peer-reviewed indexed international journals. His current research areas include innovation policies, entrepreneurship, software development, and decision support systems.

José Duarte Santos received his PhD in Management from the University of Vigo, Spain. He is also Master of Marketing and has a Bachelor degree in Business Sciences. Additionally, he obtained the title Specialist in Marketing and Advertising in accordance with Portuguese Decree-Law No. 206/August 31, 2009. Between 1987 and 2002, he has played various roles in sales, marketing, and management of companies in the information technologies sector. From 2003, he has performed functions of management and marketing consultant. Since 1999, he has been a professor in higher education in Portugal in the field of management and marketing. He is currently a professor at the Instituto Superior Politécnico Gaya (ISPGaya) and at the Instituto Superior de Contabilidade e Administração do Porto (ISCAP). He is also research at the Centre for Organisational and Social Studies of Polytechnic of Porto (CEOS.PP). His current research areas include marketing policies, social marketing, and innovation.